A C S   S Y M P O S I U M   S E R I E S   **570**

Vitrinite Reflectance as a Maturity Parameter

Applications and Limitations

Prasanta K. Mukhopadhyay, EDITOR
Global Geoenergy Research Ltd.

Wallace G. Dow, EDITOR
DGSI

Developed from a symposium sponsored
by the Division of Geochemistry, Inc.,
at the 206th National Meeting
of the American Chemical Society,
Chicago, Illinois,
August 22–27, 1993

American Chemical Society, Washington, DC 1994

Seplae
Chem

Library of Congress Cataloging-in-Publication Data

Vitrinite reflectance as a maturity parameter: applications and limitations / Prasanta K. Mukhopadhyay, Wallace G. Dow.

p.   cm.—(ACS symposium series, ISSN 0097–6156; 570)

"Developed from a symposium sponsored by the Division of Geochemistry, Inc., at the 206th National Meeting of the American Chemical Society, Chicago, Illinois, August 22–27, 1993."

Includes bibliographical references and indexes.

ISBN 0–8412–2994–5

1. Coal—Analysis—Congresses.

I. Mukhopadhyay, Prasanta K. II. Dow, Wallace G. III. American Chemical Society. Division of Geochemistry, Inc.. IV. American Chemical Society. Meeting (206th: 1993: Chicago, Ill..) V. Series.

TP325.V55   1994
553.2′4′0287—dc20   94–34670
                                                       CIP

The paper used in this publication meets the minimum requirements of American National Standard for Information Sciences—Permanence of Paper for Printed Library Materials, ANSI Z39.48–1984. ∞

Foreword

THE ACS SYMPOSIUM SERIES was first published in 1974 to provide a mechanism for publishing symposia quickly in book form. The purpose of this series is to publish comprehensive books developed from symposia, which are usually "snapshots in time" of the current research being done on a topic, plus some review material on the topic. For this reason, it is necessary that the papers be published as quickly as possible.

Before a symposium-based book is put under contract, the proposed table of contents is reviewed for appropriateness to the topic and for comprehensiveness of the collection. Some papers are excluded at this point, and others are added to round out the scope of the volume. In addition, a draft of each paper is peer-reviewed prior to final acceptance or rejection. This anonymous review process is supervised by the organizer(s) of the symposium, who become the editor(s) of the book. The authors then revise their papers according to the recommendations of both the reviewers and the editors, prepare camera-ready copy, and submit the final papers to the editors, who check that all necessary revisions have been made.

As a rule, only original research papers and original review papers are included in the volumes. Verbatim reproductions of previously published papers are not accepted.

M. Joan Comstock
Series Editor

Contents

Preface

VITRINITE REFLECTANCE IS A METHOD used to measure the intensity of reflected light from a lignocellulose component (the maceral *vitrinite*) in sedimentary organic matter. It is extremely useful for measuring the heat flux of the earth's crust through geologic time, delineating the timing of oil and gas generation, and predicting coke stability.

Developed by the European coal petrographers during the 1930s, vitrinite reflectance was first used to define the rank of coal. It has been used extensively since about 1970 by petroleum geologists, organic geochemists, and organic petrographers to solve other geological problems. When vitrinite reflectance was applied to various geological problems, its limitations were observed.

During the past 60 years, our knowledge of the applications and limitations of vitrinite reflectance expanded so greatly that this comprehensive volume could be generated. The earlier research on vitrinite reflectance was published in various journals (especially in proceeding volumes of the Society of Organic Petrology), which are included as references in the overview chapter of this book. This volume is the first of its kind and takes an up-to-date and integrated approach to combine the applications and limitations of vitrinite reflectance. Naturally, it is impossible to illustrate all aspects of vitrinite reflectance. This volume focuses on four aspects: (1) petrographic characteristics and limitations, (2) molecular characterization, (3) implications for basin modeling, and (4) correlation with other microscopic and chemical maturity parameters.

Most of the significant papers presented at the symposium on which this book is based are contained in this volume. Sixteen chapters, plus a comprehensive overview chapter, written by well-known professionals clarify the complexities of the vitrinite reflectance techniques. In anticipation of the increasing knowledge and activities around the world, we direct this book to the academic, industrial, and governmental groups of the international scientific community who deal with coal and hydrocarbon research and resource evaluation.

Acknowledgments

Acknowledgments are made to the donors of the Petroleum Research Fund, administered by the American Chemical Society, for partial support of this symposium. We also acknowledge the help of the authors who contributed to various chapters in the book, the members of the Books Department of the American Chemical Society (especially to Rhonda Bitterli and Charlotte McNaughton), and the members of the Division of Geochemistry, Inc., of the American Chemical Society (especially Pat Hatcher, Mary Sohn, Paul Philp, and Royston Filby) for both the symposium and the preparation of this volume. We acknowledge Global Geoenergy Research Ltd. and DGSI for their help with the symposium and the volume. Finally, we express our earnest gratitude to the reviewers who devoted their valuable time to reviewing critically and improving chapters of the book.

PRASANTA K. MUKHOPADHYAY (MUKI)
Global Geoenergy Research Ltd.
P.O. Box 9469, Station A
Halifax, Nova Scotia B3K 5S3
Canada

WALLACE G. DOW
DGSI
8701 New Trails Drive
The Woodlands, TX 77381

July 11, 1994

Chapter 1

Vitrinite Reflectance as Maturity Parameter

Petrographic and Molecular Characterization and Its Applications to Basin Modeling

P. K. Mukhopadhyay

Global Geoenergy Research Limited, 14 Crescent Plateau, Halifax, Nova Scotia B3M 2V6, Canada

Huminite/Vitrinite, primarily derived from the lignin, cellulose, and tannins of vascular plants (especially from the periderm [bark] and xylem [wood] tissues), is the major maceral (*1*) in coal but also occurs as dispersed organic particles in various sedimentary rocks, some of which are petroleum source rocks. The maceral "huminite" (a low rank counterpart of vitrinite) is formed when the biopolymers of plants are subjected to physical and chemical alteration due to increases in temperature and pressure through geological time. With increasing temperature and pressure within the earth's crust, irreversible chemical reactions take place within the huminite/vitrinite network. This process is termed maturation or coalification with subdivisions like diagenesis, catagenesis, and metagenesis, which eventually merge to low grade metamorphism (*2,3*). The major physical manifestation of the maturation process is the increase in the reflectance of vitrinite which may include chemical transformations such as decarboxylation, dehydroxylation, demethylation, aromatization, and finally polycondensation of aromatic lamellae (*3–6*). The reflectance of vitrinite is defined as the proportion of normal incident light reflected by a plane polished surface of vitrinite, which changes according to the level of maturation.

Vitrinite reflectance, the major maturity parameter, accurately establishes the effective maximum paleotemperature and its duration at any stage in geological time. Accordingly, vitrinite reflectance is often referred to as a *paleogeothermometer* (*7,8*). The determination of maturity is the major building block for the determination of the boundary conditions of liquid and gaseous hydrocarbon generation, coke stability prediction in a coal, identification of various geological phenomena (faulting, thrusting, intrusion, unconformity, etc.) and temperature history in basin evolution (*3,9–23*).

The reasons for choosing vitrinite as the main parameter for the determination of maturity are: (a) ubiquitous presence of vitrinite or vitrinite–like macerals in almost every organic–lean and organic–rich sedimentary or metasedimentary rock; (b) vitrinite or vitrinite–like macerals appear homogenous when viewed under the incident light microscope (in most cases, vitrinite grains are large enough for maturity

0097–6156/94/0570–0001$08.72/0

determination); and (c) vitrinite shows uniform physical and chemical changes under increasing thermal stress and other geological conditions. The objectives of this review are to illustrate the physical and chemical complexity of vitrinite reflectance, its limitations, and its application in various geological situations. In this review, a complete assessment of the petrography and chemistry of vitrinite reflectance will be undertaken in order to establish the implications of various chapters presented in this book. Four aspects of vitrinite reflectance are discussed: (a) standardization of vitrinite reflectance and problems as seen during petrographic characterization, (b) the molecular characterization of vitrinite reflectance, (c) application of vitrinite reflectance to basin modeling, and (d) other maturation parameters which can substitute for vitrinite reflectance measurement. The chapters of this book were arranged according to their relationship to those four aspects of vitrinite reflectance.

Petrography of Vitrinite: Standardization, Diversity and Limitations

Vitrinite reflectance evolved from a simple tool to determine the rank of the coal, to extended applications such as calibration criteria for various physical and chemical rank parameters which were developed later. During the 1970's, vitrinite reflectance measurements (especially in the dispersed organic matter) were oversimplified (e.g. auto imaging techniques or measurement on kerogen smear slide) which created major problems in interpreting various geological situations, such as suppression of vitrinite reflectance. Organic petrographers and geochemists, in recent years, have set forth a new research goal to solve those problems. The following section will shed some light on those problems starting with the history of vitrinite reflectance.

History and Standardization of Vitrinite Reflectance. Vitrinite reflectance is measured in both coal and dispersed organic matter (DOM) of petroleum source rocks. However, the reflectance measurement was first developed by the coal petrographers as a rank parameter for coal. In 1930, E. Stach from Germany developed the initial idea about the implications of the reflectance of various macerals in coal from the Ruhr Coalfield (24–26). Hoffmann and Jenkner (27) began measuring vitrinite reflectance with a Berek Photometer and oil immersion objective to show its correlation with other rank parameters (16,24–26). In the early days of vitrinite reflectance, some controversies existed over the uniform increase in vitrinite reflectance with the increase in rank (25). In the following years, the works of McCartney (28), Mukherjee (29), Huntjens and van Krevelen (30), and Siever (31) documented the importance of vitrinite reflectance as a tool for rank and coke prediction (16). The beginning of the modern method of vitrinite reflectance measurement is marked by E. Stach who first used a single cell photomultiplier to measure vitrinite reflectance (16).

The earliest known reflectance measurement on dispersed vitrinite in other sedimentary rocks is derived from the works of M. and R. Teichmuller in 1951–52 using whole rock polished plugs (Teichmuller, M., Krefeld, Germany, personal communication, 1993). The present day measurement of vitrinite reflectance on kerogen concentrate began during the middle of the 1960's by organic petrographers simultaneously in various oil company laboratories and other research organizations in Europe and USA (C. Gutjahr, B. Parks and J. Castano at Shell Development

Company, Netherlands and USA, J. Burgess at Exxon Production Research and Chevron Oil, USA, P. Robert at Elf Acquitaine, France) (Castano, J., DGSI, The Woodlands, Texas, personal communication, 1993). At the same time, vitrinite reflectance on dispersed organic matter was also done by M. Teichmuller (Geologishes Landesamt, Germany) and H. Jacob (BGR, Germany), B. Alpern (University of Orleans, France), I. Ammosov (University of Moskow, U. S. S. R.), P. Hacquebard (GSC, Canada) who usually used a polished whole rock or handpicked coal from a sedimentary rock. Extensive reviews of vitrinite reflectance and its applications are well illustrated in the following literatures: (a) for both dispersed organic matter and coal, see Robert (*18*), Teichmuller (*3, 36-38*), Durand et al. (*19*), and Mukhopadhyay (*21*); (b) for coal, see McCartney and Teichmuller (*32*), Berry et al. (*33*), Davis (*34*), Ting (*35*), Stach et al (*16*), and Bustin et al (*39*); and (c) for only dispersed organic matter, see Dow (*13*), van Gijzel (*15*), Whelan and Thompson-Rizer (*22*), and Senftle et al (*23*).

The reflectance system of today includes an incident light microscope, a photomultiplier, a microprocessor, and a computer. Vitrinite reflectance measurements are made on polished coal, kerogen, and rock fragments using a 25–50 power incident light oil immersion objective, immersion oil (density 1.515 gm/cm^3), a 546 nm filter, and a measuring aperture of 2–15 µm, with or without polarizer, and a standard (usually glass) with known reflectance under oil immersion objective. The standard measuring technique for vitrinite reflectance in coal is illustrated in ASTM (*40*) and for vitrinite reflectance in dispersed organic matter in Mukhopadhyay (*21*) and Whelan and Thompson-Rizer (*22*).

The linearity of the microscope is usually calibrated using standards (such as glass, sapphire, etc.) whose refractive index at 546 nm is predetermined. Before beginning vitrinite reflectance measurement, the following five criteria should be critically reviewed (*21*): (a) the polishing of the vitrinite grains should be scratch-free and relief-free, (b) the standard should be cleaned and kept in a dust-free, temperature controlled place (the standard should be calibrated periodically during the actual measurement), (c) proper and uniform vitrinite grains should be selected, (d) the proper objective and measuring diaphragm should be accurately chosen, and (e) all instrument parts should be checked and calibrated.

Usually 25 to 100 vitrinite grains are measured with a mean reflectance value calculated together with a standard deviation of the measured grains. Two types of reflectance measurements are common to petrographers: (1) mean random reflectance which is measured without any polarizer ($R_{o \, random}$, R_o or R_m) (*16, 21*); and (2) mean maximum and minimum reflectance which uses a polarizer in the light path (R_{max} or R_{min}) (*16,34,35,39*). The reflectance is either measured by rotating the microscope stage up to 360° (*16,34,35,39*) or by rotating the polarizer through 360° (*16, 34,35,39, 41*).

One chapter in this volume (*DeVanney and Stanton*) is devoted to the standardization of vitrinite reflectance. They illustrate the importance of standardization citing the examples of interlaboratory correlation through the ASTM. They suggest that the acceptable level of variation in measured vitrinite between various laboratories should be not more than 0.02% for coal when the laboratories use similar standard procedures. In their experiment, they observed that the variation of mean vitrinite reflectance in most laboratories is around 0.05%. However, some laboratories show a wide variation of up to 0.15%. Their recommendations include a strict guideline of standardization using more interlaboratory round robin studies of various coals.

Diversity of Vitrinite Macerals and Limitations of Vitrinite Reflectance. Huminite or vitrinite is not a single homogenous maceral (*1,16*). It is a group of macerals which have diverse morphological characteristics. The diversity of these macerals is controlled by various factors such as rank, plant types, etc. In coal, the greatest diversity of huminite/vitrinite macerals is seen at low maturities such as the peat or lignite stage. They are eventually homogenized as the rank increases to the medium volatile bituminous stage. The details on the complexity and diagenetic changes related to the transformation of plant biopolymers (lignin, cellulose, and tannin) to huminite and vitrinite macerals were documented in Stach et al. (*16*), Teichmuller (*3*), Mukhopadhyay (*21*), Stout and Spackman (*42*) and Mukhopadhyay and Hatcher (*43*). Beyond the low volatile bituminous stage, it is extremely difficult to determine the diversity of various vitrinite macerals petrographically.

At the coal rank between lignite to bituminous, huminite/vitrinite is divided into the macerals telinite, collinite, and detrinite on the basis of morphology, diagenetic characteritics, and texture. Table 1 shows the relationship between those macerals and their low rank counterparts. For the definition of various vitrinite macerals, see Teichmuller (*3*), Stach et al. (*16*), Mukhopadhyay (*21*) and Mukhopadhyay and Hatcher (*43*).

Telinite macerals show remnant plant cell structures, whereas *collinite* macerals have amorphous appearances and the *detrinite* macerals have fragmented characters which are sometimes partially gelified. Collinite macerals in the subbituminous and bituminous rank, have four maceral types: telocollinite, gelocollinite, desmocollinite and corpocollinite. Gelocollinite can occur either as pore–filling within other vitrinite submacerals or occur as large individual bands. The large, amorphous individual bands which are totally corroded by etching are called *gelinite (44)*. The reflectance of telocollinite, gelocollinite and corpocollinite are usually higher than desmocollinite, which contains a fine liptinitic matrix. Another gelified vitrinite maceral which shows lower (at least <0.2% R_o) reflectance than the corresponding desmocollinite, is termed *saprocollinite (21)* and is recognized as part of the vitrinite group. Saprocollinite which is considered as a "vitrinite with suppressed reflectance", is often associated with sapropelic coal, oil shale, and within dispersed organic matter in Kerogen Type I and II (*21*). For a standard vitrinite reflectance measurement of coal and dispersed organic matter in a petroleum source rock, telocollinite grains are usually chosen. However, sometimes even in a humic coal, the proper identification of telocollinite grains can be difficult.

The chapter by <u>Bensley and Crelling</u> in this volume suggests that the heterogeniety within a single vitrinite maceral can be the cause for the scattering of data points in mean vitrinite reflectance measurement; the heterogeniety within a single telocollinite (vitrinite) can be evaluated by an etching technique (*43,44*). Accordingly, the remnant cell structures (the telinite part) within a telocollinite grain shows a higher reflectance compared to the etched areas (the gelinite part; *44*). They also demonstrate that increasing the measuring aperture (from 2 to 10 μm) converges the heterogeniety and results in a tighter reflectance histogram. This will show the

Table 1: Classification of vitrinite macerals in coal and dispersed organic matter [modified after Stach *et al.* (*16*) and Mukhopadhyay (*21*).

Low Rank (R$_o$ < 0.55%)				High Rank (R$_o$ > 0.55%)		
Maceral Group	Maceral Subgroup	Maceral	Maceral Type	Maceral Type	Maceral	Maceral group
Huminite	Humotelinite	Textinite		Telinite 1	Telinite	Vitrinite
		Ulminite	Texto–ulminite	Telinite 2		
			Eu–ulminite			
	Humocollinite	Corpohuminite	Phlobaphinite	Corpocollinite		
			Pseudo–Phlobaphinite			
		Gelinite	Levigelinite	Telocollinite	Collinite	
			Porigelinite	Gelinite		
				Gelocollinite		
			Saprohuminite (*)	Saprocollinite (*)		
	Humodetrinite	Densinite		Desmocollinite	Detrinite	
		Attrinite		Vitrodetrinite		

(*) = These macerals represent suppressed vitrinite reflectance

genuine heterogeniety between the telinite and gelinite parts within a single grain of so-called "telocollinite".

The choice of huminite/vitrinite grain in a dispersed organic matter (especially in a marine anoxic environment) of a Kerogen Type I and II source rock is more complex. In sedimentary rocks other than coal, vitrinite macerals are present as either autochthonous or allochthonous grains. The allochthonous grains are derived from the erosion of older rocks which are incorporated within the autochthonous population. The other causes of variation of vitrinite macerals in dysoxic to anoxic lacustrine, deltaic, and marine environments are: (a) a higher pH which enhances gelification and the incorporation of lipids within the lignin components by bacteria and other scavangers, (b) the low diversity of plants and abundance of phytophanktons, and the lower content of organic carbon in a marine environment (compared to peat-forming areas), unless it is affected by a high influx of terrestrial input (45), and (c) the lack of woody vegetation in the terrestrial environment, whereas algal and marsh vegetation are more common.

In a lacustrine, backswamp and upper delta plain environment with a dysoxic to anoxic conditions, the proportions of vitrinite submacerals vary widely when a humic coal is compared with a sapropelic coal (3, 16, 21-23, 45-49). In a humic coal (presumably derived mainly from arboreal vegetation), the prevalent macerals are telocollinite, gelocollinite, and desmocollinite within the vitrinite population which consists of about 50-90% of the organic constituents. On the other hand, in a sapropelic coal (presumably derived mainly from marsh and aquatic plants), corpocollinite and saprocollinite are the major macerals which make up about 10-40% of the coal (21-23, 46-49). In a lower delta plain environment, desmocollinite (presumably derived from marsh plants) dominates over telocollinite. In a shallow marine shale or limestone with abundant terrestrial input, desmocollinite and gelocollinite macerals predominate because of active gelification due to the high pH at the sediment-water interface; abundant second cycle (recycled) vitrinites are also present (21-23). In a shale or limestone formed in a highly anoxic environment, vitrinite macerals are extremely rare and whenever present mainly consist of corpocollinite and saprocollinite derived partly from the degradation of marine grasses, algal cells or transported herbaceous/arboreal vegetation (21,22,45,46,50-52).

In pre-Devonian rocks where vascular land plants are scarce to absent, vitrinite-like macerals are mainly derived from the polysaccharides components of the algal cells (21-23,50,52). Similarly, akinete cells of cyanophytes (algae) can produce corpohuminite-like macerals as discussed in the chapter by Stasiuk et al. They demonstrate that the reflectance of corpohuminite (0.2-0.8% R_o) derived from algae and their associated T_{max} (412 to 445°C) values correlate with vitrinite (derived from the arboreal vegetation) reflectance in Devonian and Mississippian rocks from Saskatchewan and Alberta, Canada. This correlation is extremely important as the thermal maturity of pre-Devonian rocks from basins elsewhere can be properly evaluated through measurement of algal-derived vitrinite surrogates.

For petrographic identification, the vitrinite submacerals in coal are often large enough for easy identification whereas vitrinite macerals in dispersed organic matter occur mostly as tiny grains and often show suppressed reflectance. Apart from the extreme variation of primary vitrinite populations, secondary macerals (example: solid bitumen) can cause problems in identifying an autochthonous vitrinite population because they show close similarity morphologically (*21-23*). The morphological and comparative characteristics of vitrinite macerals within a bituminous coal and vitrinites or solid bitumen in dispersed organic matter within an immature and a mature source rock are illustrated in Figures 1A-D as an example. Figure 1A shows a large homogenized grain of telocollinite and Figure 1B has several medium-sized grains of telocollinite, desmocollinite, and gelocollinite along with inertodetrinite and liptinite macerals. In contrast, Figures 1C and 1D point out the tiny grains of collinite within the matrix of amorphous liptinite (AOM 2; *45*) from a typical Kerogen Type II oil-prone source rock. The solid bitumens in Figure 1D show a close similarity in properties with collinite macerals. In this case, they are differentiated only by their reflectance. The problem associated with the identification of vitrinites in a sedimentary rock cuttings recovered from boreholes and which are contaminated by drilling mud additives or caving during drilling were documented earlier by Dow (*13*), Mukhopadhyay (*21*), and Senftle et al. (*23*).

Vitrinite reflectance measured on whole rock is often considered to be more reliable because autochthonous vitrinite grains are better identified in their natural setting compared to the vitrinites in isolated kerogens (*3,21*). However, the measurement of vitrinite reflectance in a whole rock is often extremely time-consuming and show lower (0.05-0.25% R_o) values compared to the measured vitrinites in an isolated kerogen polished plug (*46,53*). The lowering of vitrinite reflectance in a whole rock polished plug is possibly caused either by the impregnation of bitumen or the non-availability of telocollinite grains which are due to the variation in organic facies. In an organic- or vitrinite-lean rock, the measurement of vitrinite reflectance using isolated kerogen is recommended.

Suppression of Vitrinite Reflectance. Within the *oil window*, the reflectance of vitrinite in marine shales or lacustrine coal and shales is often much lower (0.15% to 0.55%) in comparison to humic coals which may overlie or underlie the shales. This suppression of vitrinite reflectance has been previously documented by many workers (*54-63*). As discussed by Mukhopadhyay (*21*), the suppression or lowering of vitrinite reflectance in a coal or in dispersed organic matter is caused by: (1) lithological variation and differences in thermal conductivity and heat capacity (*21* and Gentzis and Goodarzi, this volume), (2) the formation of perhydrous vitrinite or vitrinite-like macerals due to lipid incorporation within the biopolymers derived from lignin, cellulose or tannin. This lipid incorporation is possibly caused by variation in pH at the sediment-water interface or variation in chemistry within various plant components (*64*) or the formation of a vitrinite-like maceral from algae, marine grasses, etc. (*21,46,50,52,56* and Stasiuk, this volume), (3) the impregnation of generated bitumen or incorporation of migrated oil (Suarez-Ruiz et al., this volume), (4) the abundance of liptinite macerals in a vitrinite-poor source rock (*54*), (5) the variable degrees of bacterial activity in the sediment (Quick, this volume), and (6) sample contamination due to cavings derived from the younger horizons above or mixing with lignitic drilling mud additives (*13, 21, 23*).

In this volume, three chapters by Quick, Suarez-Ruiz *et al.*, and Gentzis and Goodarzi, are solely devoted to the suppression of vitrinite reflectance. A part of the

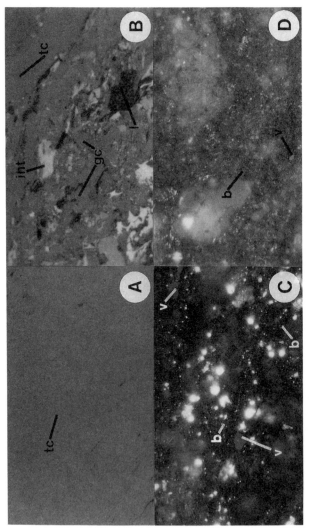

Figure 1. Photomicrographs in incident white light under oil immersion objective. X450. (A) Telocollinite (tc) in a bituminous coal, Nova Scotia, Canada. R_o = 0.75%, (B) telocollinite (tc), gelocollinite (gc), desmocollinite, vitrodetrinite, inertodetrinite (int), and liptinite (l) in a bituminous coal, Nova Scotia, Canada. R_o =0.88%, (C) Huminite/vitrinite (v) within a matrix of amorphous liptinite and solid bitumen (b), Devonian shale (Kerogen Type II), USA. R_o = 0.45%, and (D) vitrinite (v) and solid bitumen (b) within a matrix of amorphous liptinite, Cretaceous La Luna shale, Venezuela (matured Kerogen Type II). R_o = 0.9%.

chapters written by Veld *et al*. and Hill *et al.* also focus on this topic. Quick, working with New Zealand coals (0.5 to 1.1% R_o), suggests that the suppression of vitrinite reflectance in those coals is caused by sulphate–reducing bacteria forming hydrogen–rich vitrinite macerals which show lower reflectance and higher fluorescence. He shows negative correlation between the abundance of sulphur (in weight percent) with vitrinite reflectance and positive correlation with fluorescence intensity. Accordingly, the suppression of vitrinite can be calibrated with the measured fluorescence intensity of vitrinites from the same coals.

In their study on the geochemistry of Spanish perhydrous vitrinite (bitumen impregnated, i.e., *Jet*) bearing coals, Suarez–Ruiz *et al.* also demonstrate the suppression of vitrinite reflectance. It shows an anomalous variation of huminite/vitrinite reflectance within two huminite/vitrinite macerals in the same coal (0.39% for ulminite [telocollinite, Table 1] and 0.72% for phlobaphinite [corpocollinite, Table 1]) and extremely low reflectance (0.39% for the ulminite) compared to the carbon content (85%). The ulminite (making up 85% of the *Jet*) has a higher fluorescence intensity compared to phlobaphinite (comprising 15% of the *Jet*) which is nonfluorescent. Those perhydrous coals (which have a high hydrogen index and bitumen extract) when analysed via Fourier Transformation Infrared Spectroscopy (FTIR), showed accumulation and absorption of migrated oil within their ulminite cell structures.

Gentzis and Goodarzi also illustrate the suppression of vitrinite reflectance. Their study examines the maturation profile of Mannville Group coals from Western Canada. Using Rock–Eval pyrolysis, biomarker GC–MS analyses, and quantitative fluorescence measurements, they suggested two possibilities for the cause of vitrinite reflectance suppression: (a) the development of perhydrous, fluorescent vitrinite in a brackish water environment as revealed by high sulphur and boron content; or (b) the simultaneous generation of liquid hydrocarbons from the perhydrous vitrinites and the impregnation of migrated oil within the vitrinite network.

Calibration of Suppressed Vitrinite Reflectance. During the 1990s, organic petrographers and geochemists have put forward suggestions for the correction of suppressed vitrinite (saprocollinite) reflectance to the level of standard vitrinite reflectance of a telocollinite grain. Quick (this volume) suggested the calibration of suppressed and nonsuppressed vitrinite macerals with fluorescence intensity; (b) Wilkins et al. (*63*) used fluorescence alteration of various macerals to evaluate the suppressed and nonsuppressed vitrinite.

The following are the two most significant and simple calibration methods to define the suppressed and nonsuppressed trends using hydrocarbon potential of the source rock. Lewan (*65*), using hydrous pyrolysis experimentation, demonstrated the parallel reflectance trends of suppressed and nonsuppressed vitrinites in marine and lacustrine shales compared to humic coal (Figure 2). The maturation trend of humic coal is considered as the nonsuppressed true vitrinite reflectance trend. Lo (*66*) employed the hydrogen index (derived from Rock–Eval pyrolysis) of suppressed and nonsuppressed shales and coal as a correction factor for the perhydrous vitrinite (Figure 3). Accordingly, a source rock will show progressive decrease in vitrinite reflectance as the hydrogen index increases from 100 to 950 mg HC/g TOC. The

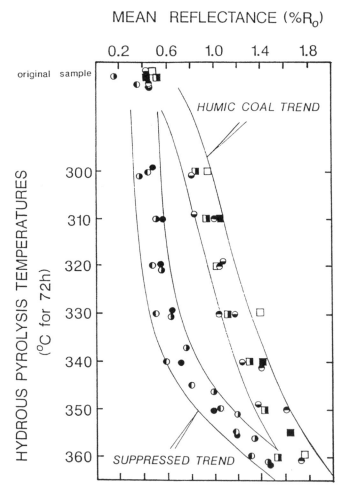

Figure 2. Vitrinite reflectance suppression model showing suppressed (dispersed organic matter in Kerogen Type II) and nonsuppressed (humic coal, Kerogen Type III) trends (after Lewan, 65). The symbols for various trends are: humic coal – open squares = Blackhawk Fm., half solid squares = Frontier Fm., solid squares = Wilcox Fm.; shales – right half solid circles = Phosphoria Fm., left half solid circles = Woodford Shale, solid circles = Alum Shale, bottom and top half solid circles = Mowry Shale. All samples are heated isothermally at temperatures from 300° to 360°C in contact with liquid water for 72 hour durations.

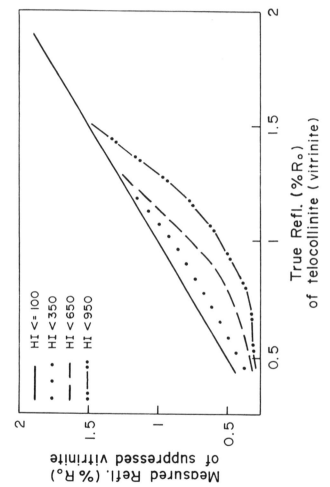

Figure 3. Preliminary model for the correction of the suppression for vitrinite reflectance trends using the amount of hydrogen index (mg HC/g TOC from Rock–Eval pyrolysis) as major criteria for the suppression of vitrinite reflectance (modified after Lo, 66).

suppressed reflectance can be calibrated to the reflectance of telocollinite which is considered as nonsuppressed true vitrinite reflectance. Telocollinite or gelocollinite maceral is usually less than or equivalent to 100–150 mg HC/g TOC (*47, 67*).

Molecular Characterization of Vitrinite Reflectance

The increase in huminite/vitrinite reflectance from the peat to anthracite stage is controlled by macromolecular changes of lignin biopolymer (vitrinite precursor) during coalification. Some recent reviews have discussed these important changes (*3,5,6,19,21*). During early diagenesis, only a minor fraction of biopolymers, such as cellulose and tannins, are preserved because they are involved in hydrolysis and oxidation by fungi and bacteria. The lignin biopolymers, on the other hand, are subjected to a series of defunctionalization reactions during the formation of huminite (vitrinite precursor) (*3,5,6,43,68*). Huminites are eventually transformed to the macerals of the *vitrinite* group around the subbituminous coal rank through a series of sequential reactions (*5,6*). During the early catagenetic stage, the process of dehydroxylation and decarboxylation of phenolic compounds forms benzofurans, alkyl benzenes, and alkyl naphthalenes through a cleavage of alkyl–oxygen bonds in aryl–alkyl ethers (*5,6,21,69,70*). Figure 4 illustrates the molecular changes that occur within the structural network of vitrinite macerals starting from the biopolymer lignin to the formation of vitrinite at the high volatile bituminous stage (*6*). The increase in vitrinite reflectance in the late catagenetic and metagenetic stages are mainly caused by the polycondensation of the aromatic compounds which are eventually oriented to aromatic clusters due to load pressure (*3,4,16,21,43*). The complexity of molecular changes within the vitrinite group with increasing reflectance up to the medium volatile bituminous rank may be better evaluated with the help of Mico–FTIR, Micro–Laser pyrolysis GC–MS, and NMR–Microscopy. Some of the recent research using those techniques laid the foundation for the molecular characterization of vitrinite reflectance (*6,71–73*).

Five chapters of this volume (Hatcher *et al.*, Kruge and Bensley, Veld *et al.*, Hill *et al.*, and Laggoun–Defarge *et al.*) discuss the complexity of molecular changes associated with vitrinite reflectance through the rank succession of peat to anthracite. One other paper in this volume (Sackett *et al.*) deals with the molecular changes of CH_4 and CO_2 in a marine source rock in which they develop a new chemical maturation parameter that is correlatable with vitrinite reflectance.

Using hydrous pyrolysis and NMR, Hatcher *et al.* simulated the chemical maturation reactions in coal from the peat to bituminous stage. The reactions involved hydrolysis of aryl ethers, reduction of alkoxyl side chains, alkylation of reactive aromatic sites and, finally, the condensation reactions of phenolic moieties. The hydrous pyrolysis experimentation induced the transformation of catachol–like structures to phenol–like structures. However, in this study, the reactions related to the aliphatic structures in coal do not represent the natural processes. The reactions took thermolytic pathways instead of ionic pathways. The alkyl radicals possibly penetrated the aromatic ring sites and induced alkylation with an increase in the average number of aromatic substituents.

Kruge and Bensley utilized pure vitrinite concentrates from Appalachian

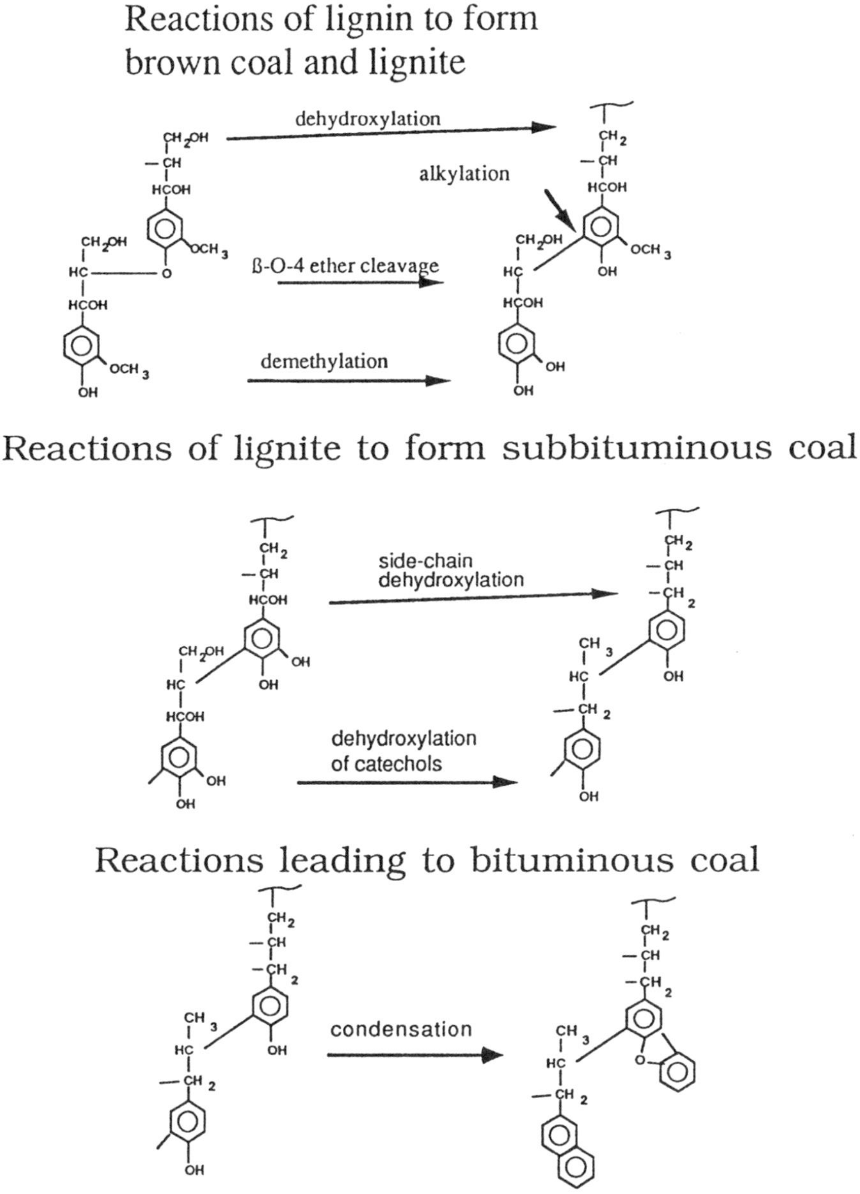

Reactions of lignin to form brown coal and lignite

Reactions of lignite to form subbituminous coal

Reactions leading to bituminous coal

Figure 4. Stages of reaction pathways for the chemical transformation of lignin to vitrinite (after Hatcher, 6).

Carboniferous coals (rank between 0.66 to 1.39% R_{max}) for flash pyrolysis–GC/MS studies to illustrate the changes in the aromatic and hydroxy–aromatic blocks within the coal network. The relative concentration of tri– and tetra–aromatic hydrocarbons (especially benzo [α] fluorene, methylphenanthrenes, methyl fluorene, and methyl chrysene isomers) exhibit marked increase around 0.9% R_{max} with a systematic loss of phenolic compounds. These results demonstrated that some individual isomers of some polyaromatic hydrocarbons (PAH's) change properties as the rank increases during pyrolysis. Two of the isomer ratios 1,3,7–,13– and 2,3,6–tetramethyl naphthalenes and 2,6– and 2,7– dimethyl phenanthrenes show systematic changes. The ratios between chrysene (higher stability to temperature) and benzo (α) anthracene or benzo (α) fluorene and methylpyrenes show a sharp increase in concentration between 0.82 and 1.02% R_{max} of vitrinite concentrates. This increase coincides exactly with the "Second Coalification Jump" as defined by the coal petrographers (Stach et al., 1982). Stach et al. (16) suggested that, at this stage, the manifestation involving physical changes in vitrinite (the apparent increase in fluorescence), runs parallel to the chemical changes including the overall increase in tri– and tetra–aromatic hydrocarbons in vitrinites.

Veld et al. use a similar type of pure vitrinite concentrate from the Carboniferous coals of the Ruhr area, Germany (reflectance: 0.96 to 1.93% R_{max}) to pyrolyze under Curie Point Flash Pyrolysis (temperature: 780°C). The concentration of four major products (alkylbenzene, alkylphenol, naphthalene and phenanthrene) were quantified using GC–MS. Alkylphenanthrene and alkylbiphenyls increased significantly with increased maturation (especially between 0.96 to 1.35 %R_{max} – near the "Second Coalification Jump" of liptinites), whereas alkylphenol, alkylbenzene, and alkylnaphthalene decreased. The presence of benzene and naphthalenes in the original coal, which are not usually present in the lignin pyrolysis products, suggest that the biopolymer within the lignin–rich macromolecule was incorporated during early diagenesis. Vitrinite reflectance measurements of one unextracted and extracted pyrolyzed coal sample (rich in bitumen) reveals a significant difference of 0.16% R_{max} which suggest vitrinite reflectance suppression in the nonextracted sample. Veld et al. propose that the suppression of vitrinite reflectance is caused by the absorption of significant fractions of photons to the incident light beam by the aliphatic components present in the nonextracted samples.

Hill et al. demonstrate the importance of pressure in the increase or decrease of vitrinite reflectance and the generation of CH_4 and CO_2 using a gold tube hydrous pyrolysis experiment on a bituminous coal. The rates of vitrinite maturation and CH_4 or CO_2 generation increased when the experimental pressure approaches 600 bar. The vitrinite reflectance increase to 1.01% R_o (at 300°C) and to 1.56% R_o (at 340°C) from a starting coal of 0.5% R_o. Beyond pressures of 600 bar but less than 2000 bar, the gas generation and vitrinite reflectance is retarded to 0.83% (at 300°C with 2000 bar) and 1.43% R_o (at 340°C with 2000 bar), respectively. They also show increased vitrinite reflectance suppression with increasing bitumen impregnation within the telocollinite.

Laggoun–Defarge et al. correlate the physical and bulk chemical transformation (during the increase in maturity), of a relatively immature vitrinite–rich coal by using a gold tube confined pyrolysis. They experimented with six different pyrolysis temperatures (300°, 320°, 340°, 380°, 400°, and 450°C). They have correlated the changes in the coal at every pyrolysis temperature by comparing Rock–Eval pyrolysis and TEM–EDX data with X–Ray difraction results. The ultrathin sections of vitrinite grains show that thermal degradation at 320°C (R_m = 1.18%) affected the homogeneous vitrinite (gelified part) compared to the residual cell walls (telinite part) with fracturing and the addition of uranium aggregates. The peak of the plastic stage was reached at 340°C (R_m = 1.51%) which shows maximum hydrocarbon generation. The resistant cell walls dissappeared at 380°C (R_m = 2.02%). Even at 450°C (R_m = 2.71%), the ultrafine structures of vitrinite do not show any molecular orientation and appear to be only a poorly organized carbonaceous residue. X–Ray difraction analyses, however, show a lateral organization of the basic structural units in turbostatic layers.

As an alternative to vitrinite reflectance measurement in vitrinite–poor organic–rich rocks (e.g. Kerogen Type II Miocene Monterrey calcareous shale, California), Sackett et al. demonstrate that the pyrolysis–carbon isotope (PCI) method shows a uniform change in $\delta^{13}C$ values in a natural sequence. The PCI method is based on an exhaustive pyrolysis of whole rock samples; it measures the amount and carbon isotopic compositions of the pyrolysis derived methane and carbon dioxide. Pyrolysis at 700°C for one hour seemed to be most suitable because there no isotopic exchanges between generated methane and carbon dioxide have been observed at that temperature.

Application of Vitrinite Reflectance in Basin Modeling

The earliest known use of vitrinite reflectance as a calibration parameter for maturation modeling was cited by Bostick (9), who used Karweil's (74) and Lopatin's (75) time temperature curves which correlate temperature, duration of heating (in millions years), and coal rank parameters. Although Tissot (76) first introduced the basic kinetic model, the earliest known kinetic maturation modeling was documented by Lopatin (75) who assumed that the change in the time temperature index was caused by doubling the reaction rate in every 10°C increase in paleotemperature. Waples (77) modified Lopatin's time–temperature index to be geologically more useful. All of the earlier kinetic models, however, end up with very low activation energies (74,75,78) which are unrealistic for vitrinite maturation.

Since 1975, the use of a maturation model in basin analysis has become extremely common. At present, there are basically three types of basin modeling using vitrinite maturation: (a) empirical – the method uses either temperature or time/temperature index as a maturation index with little consideration of kinetic variations (7,8,77–82); (b) kinetics 1 (single reaction) – the main criteria for vitrinite and Kerogen Type II oil source rock maturation is to follow a first order chemical reaction with a single activation energy and arrhenius constant with or without the effect of time and temperature (20,83–90); (c) kinetics 2 (parallel reactions) – the method suggests that vitrinite maturation requires parallel first order chemical reactions with a distribution of activation energies. This implied a stronger influence of temperature than time (91–95). According to the *Easy R_o* method of Sweeney and Burnham (93), the distribution of activation energies are from 34 to 74 kcal/mole with a single arrhenius factor for the generation of H_2O, CO_2, CH_n, and CH_4 from vitrinite. Suzuki et al. (95) modified the *Easy R_o* method to *Simple R_o* method where the vitrinite maturation model is based on a single activation energy at each step increase which is related to the change of vitrinite weight loss ($\%R_o$).

The kinetic model of Sweeney and Burnham (93) is the most widely used concept in basin modeling for vitrinite maturation. This model would be extremely useful in deciphering the maturation trends in a depth profile of suppressed and nonsuppressed vitrinites. The *Easy R_o* method is based on the kinetics of H/C ratios in nonsuppressed telocollinite. Pyrolysis experimentation on both suppressed and nonsuppressed vitrinites in the future will bring a clear understanding to this subject. Four chapters (Barker and Pawlewicz, Mukhopadhyay et al., Arne and Zentilli, and He Wei et al) in this volume deal with the vitrinite maturation model and its correlation with other maturation parameters.

Barker and Pawlewicz compare the kinetics and vitrinite reflectance geothermometer (VRG) models, both of which use vitrinite maturation in a global system. Citing some type examples of normal basins without much structural or thermal complexities, they demonstrate that the VRG model correlates closely with the kinetic model. However, they suggest that the kinetic model (such as Sweeney and Burnham, 93) would be more refined if the heating rate can be measured. The VRG is selected to be more accurate in cases where the tectonic regime of a basin is highly complex and the T_{peak} can be accurately measured (such as in a geothermal system).

Mukhopadhyay et al. demonstrate the complexities of vitrinite maturation and oil/gas generation of the Jurassic–Cretaceous rocks from the Scotian Basin, Nova Scotia by using *BasinMod* (a kinetic basin modeling program developed by Platt River Associates (94). The *BasinMod* study utilized the *Easy R_o* kinetics of Sweeney and Burnham (93). The measured and modeled R_o were calibrated using a rifting heat flow at 200 Ma which decreases with time. Rifting heat is dependant on β–factor (the ratio of the stretching in the crust) and present day heat flow. Using the measured kinetics of various source rocks, the hydrocarbon (HC) generation and migration in the basin were evaluated. Modeled HC generation from Kerogen Type IIA–IIB or IIB source rocks indicates that they differ from vitrinite maturation kinetics (Kerogen Type III source rock). Variations in maturation profiles are predicted to be caused by abnormal heat transfer due to changes in thermal conductivities or the presence of older rocks (closer to rifting).

Arne and Zentilli demonstrate the complementary nature of vitrinite maturation and apatite fission track annealing in a real basin situation. The total annealing of fission track in detrital apatite grains usually coincides with a R_{max} of 0.7 to 0.9% which requires an effective heating rate of 10^6 to 10^8 years. They suggest that both the vitrinite maturation and apatite fission track annealing have first order arrhenius reaction kinetics which show a parallel effects on the vitrinite reflectance (organic) and the apatite fission track (inorganic). In order to resolve the thermal evolution in a structurally complex basin, the complimentary use of the apatite fisssion track thermochronology and vitrinite reflectance measurement is recommended to predict the natural system.

He Wei et al. present a numerical model which recontructs the paleoheat flux from maturation parameters like vitrinite reflectance and smectite/illite conversion.

This model utilizes various dependencies of heat flux on time (e.g. exponential, parabolic, etc.). Accordingly, the heat flux variations are not dependent on time. Similar to the *Easy R_o* method of Sweeney and Burnham (*93*), this model used activation energy distributions rather than a single reaction. This kinetic model of vitrinite maturation is valid in controlling paleoheat flux beyond 0.4% R_o.

Correlation of Vitrinite Reflectance and Other Maturation Parameters

As vitrinite reflectance is considered to be the most suitable maturation parameter for solving various geological and geochemical problems, other microscopic and chemical parameters are frequently correlated with vitrinite reflectance in order to add certainty (*2,3,14,16,21,96-97*). This brings an additional advantage to maturity determination of a rock where vitrinite macerals are scarce or of doubtful nature (e.g. Pre–Devonian rocks, Kerogen Type I and II, contaminated samples, etc.). This will also enhance our capability to define the maturity of an oil or bitumen which can be used for oil–oil and oil–source rock correlation, and timing of migration of bitumen and gas to form reservoired oil and gas.

At present, more than twenty–five other maturation parameters can be identified; some are microscopic and others are chemical parameters. The important microscopic parameters are Thermal Alteration Index (*TAI*) (*96,98,99*) measured on sporinite, λ_{max}/Q/alteration/absolute intensity (*3,15,21,63,100-109*) in fluorescence, Conodont Alteration Index (*CAI*) (*21,109,110*), Graptolite/Chitinozoan/Scolecodont Reflectance (*109-116*), Bitumen Reflectance (*110,117,118*), and Alginite Reflectance (*116*; Stasiuk, this volume). The chemical maturation parameters include both kerogen and bitumen/oil analyses. A historical review of chemical maturation parameters (bulk chemical and biomarkers) are discussed in Tissot and Welte (*2*), Waples and Machihara (*119*) and Radke (*120*). Figure 5 shows the correlation of various microscopic and chemical maturation parameters. In this figure, vitrinite reflectance is correlated with: (a) microscopic maturation parameters: TAI (*98,99*), CAI (*109*), change on alginite color (*106*), λ_{max} and Q (*3,21,100-108*), solid bitumen reflectance (*110,117,118*); (b) chemical maturation parameters: $T_{max}(°C)$ (*21,23,67,121*), Methylpenanthrene Index (MPI 1; *120,122*), Methyldibenzothiophene Ratio (*MDR*; *120*), 20S/(20S+20R) C_{29} normal steranes (*119,123,124*), the ratio between C_{27} diasteranes to total C_{27}–steranes (*114,119*), and apatite fission track annealing (Arne and Zentilli, this volume); and (c) possible zones of hydrocarbon (HC) generation (*2,21*). For details on these chemical maturation parameters, readers are referred to the literature. As discussed in various literatures (*2,119*), the biomarker ratios have limitations as they are highly susceptible to variation in organic facies, kerogen type, contamination, and migration effect (references in Waples and Machihara, *119*). Similar to vitrinite reflectance, biomarker ratios should be used with extreme caution for maturity determination. In vitrinite–poor rocks, Thermal Alteration Index (*TAI*), solid bitumen reflectance and fluorescence parameters (λ_{max}, Q, and alteration) are the most suitable microscopic maturation parameters and can be complimented with T_{max} values from the Rock–Eval pyrolysis. For crude oil and bitumen, aromatic biomarkers (MPI or MDR) are more suitable than aliphatic biomarkers as maturation parameters because they are less affected by biodegradation and migration. Three chapters in this

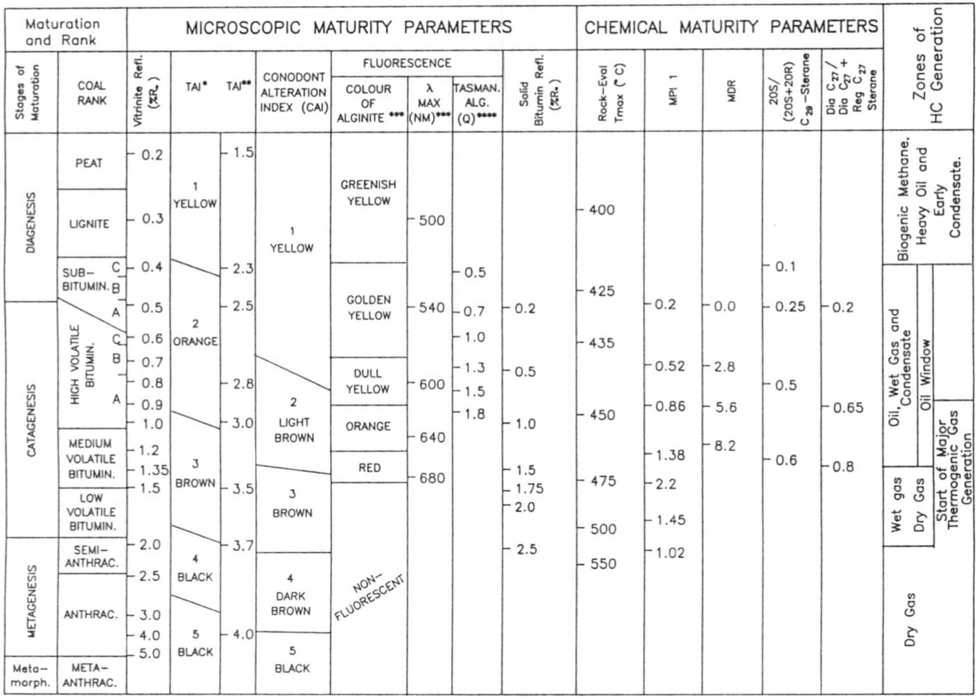

Figure 5. Correlation of microscopic and chemical maturity parameters (modified after Mukhopadhyay, 21). The symbols are with various stars indicates that the data is taken from the following sources: one star is from Staplin (98); two stars indicate from Jones and Edison (99); three stars indicate from Mukhopadhyay and Rullkotter (106); four stars indicate from Teichmuller (3).

volume (Quick, Sackett et al., and Arne and Zentilli) illustrate the correlation of fluorescence, pyrolysis carbon isotope (*PCI*), and apatite and zircon fission track annealing with vitrinite reflectance.

Conclusion

This overview chapter gives the reader a comprehensive outlook on various applications and limitations of vitrinite reflectance when it is used as the prime maturity parameter. In this review, a complete assessment of petrographic and molecular characterization of vitrinite reflectance, and use of vitrinite reflectance in basin modeling was presented in order to establish the implications of various chapters presented in this book. Four aspects of vitrinite reflectance discussed include: (a) standardization of vitrinite reflectance and problems during petrographic characterization (especially suppression of reflectance and its possible remedy), (b) the chemical characterization and molecular changes related to the increase in vitrinite reflectance including the changes from lignin biopolymer to vitrinite geopolymer, (c) aspects of vitrinite reflectance to basin modeling (especially using chemical kinetics), and (d) other maturation parameters which can substitute for the vitrinite reflectance in a vitrinite-lean rock.

Acknowledgments

The author acknowledges the help of Dr. M. H. Alimi of Global Geochemistry Corporation, California for some biomarker maturity parameter data. The author acknowledge the help of two anonymous reviewers, W. D. Smith of the Nova Scotia Department of Environment, and Dr. John H. Calder of the Nova Scotia Department of Natural Resources, Halifax, Nova Scotia for their critical review and suggestions to improve the manuscript. The author acknowledges Dr. Pat Hatcher, Pennsylvania State University, Dr. Mike Lewan, US Geological Survey, and Dr. H. B. Lo, Marybeth McAlister, Editor, *Energia,* Center for Energy Resources, Lexington, Kentucky, and Elsevier Science Publication, Amsterdam for their permission to publish some of the figures. The author also acknowledges that Figure 3 is reprinted (modified version) from the publication of H. B. Lo, *Organic Geochemistry*, vol. 20, no. 6, Copyright (1993), Figure 1, with kind permission from Elsevier Science Ltd, The Boulevard, Langford Lane, Kidlington OX5 1GB, UK.

Literature Cited

(1) International Committee of Coal Petrology. *International Handbook of Coal Petrography*, 2nd Edition, C.N.R.S. Paris, 1971.
(2) Tissot. B.; Welte, D. H. *Petroleum Formation and Occurrence*. Springer–Verlag, 1984, Second Edition.
(3) Teichmuller, M. In *Coal and Coal-Bearing Strata: Recent Advances*. Scott, A. C., Ed.; Geological Society of London, Special Publication, London, 1987a, Vol. 32, pp. 127–169.

(4) Oberlin, A.; Boulmier, J. L.; Villey, M. In *Kerogen*, Durand, B., Ed.; Editions Technip, Paris, 1980, pp. 191–242.

(5) Hatcher, P. G.; Lerch, H. E. III; Kotra, R. K.; Verheyen, T. V. *Fuel*, **1988**, *67*, 1067–1075.

(6) Hatcher, P. G. *Energia*, **1993**, *4(5)*, 1–6.

(7) Barker, C. E.; Elders, W. A. *Geothermics*, **1981**, *10*, 207–223.

(8) Price, L. C. *Jour. Petrol. Geol.* **1983**, *6*, 5–38.

(9) Bostick, N. H. *C. R. 7me Congr. Int. Stratgr. Geol. Carbonifer.* **1973**, *2*, 183–183.

(10) Burgess, J. D. *Geol. Soc. Amer. Spec. Paper,* **1974**, *153*, 19–30.

(11) Castano, J. R.; Sparks, D. M. *Geol. Soc. Amer. Spec. Paper*, **1974**, *153*, 31–52.

(12) Alpern, B. *Bull. Centre. Rech. Pau–SNPA*, **1976**, *10 (1)*, 201–220.

(13) Dow, W. G. *Jour. Geochem. Explor.* **1977**, *7*, 79–99.

(14) Bostick, N. H. *Soc. Econ. Paleontol. Mineral. Soc. Publ.* **1979**, *26*, 17–43.

(15) Van Gijzel, P. In *How to Assess Maturation and Paleotemperature*. Society of Economic Paleontologists and Mineralogists Short Note, 1981, No. 7, pp. 159–216.

(16) Stach, E.; Mackowsky, M. Th.; Taylor, G. H.; Chandra, D.; Teichmuller, R. *Stach's Textbook of Coal Petrology*, Gebrder Borntraeger, Berlin, 1982, 2nd Edition.

(17) Cardott, B. J.; Lambert, L. W. *Bull. Am. Assoc. Petrol. Geol.* **1985**, *69 (11)*, 1982–1998.

(18) Robert, P. *Historire Geothermique et Diagenese Organique.* Bull. Des. Cetres. de Recherches Expl. Prod. Elf. Aquitaine, Pau, France, 1985 (English Edition by D. Reidel, Dordrecht, Netherland, 1988).

(19) Durand, B.; Alpern, B.; Pittion, J. L.; Pradier, B. In *Thermal History of Sedimentary Basins*. Durand, B., Ed.; 1987, Technip, Paris, pp. 441–474.

(20) Tissot, B,; Pelet, R,; Ungerer, Ph. *Bull. Am. Assoc. Petrol. Geol.* **1988**, *71(12)*, 1445–1466.

(21) Mukhopadhyay, P. K. In *Diagenesis III, Developments in Sedimentology*, Wolf, K. H.; Chilingarian, G. V., Eds.; Elsevier Science Publishers, Amsterdam, 1992, Vol. 47, pp. 435–510.

(22) Whelan, J. K.; Thompson–Rizer, C. L. In *Organic Geochemistry: Principles and Applications*, Engel, M. H.; Macko, S. A., Eds., Plenum Press, New York, 1993, pp. 289–152.

(23) Senftle, J. T.; Landis, C. R.; McLaughlin, R. In *Organic Geochemistry: Principles and Applications*, Engel, M. H.; Macko, S. A., Eds., Plenum Press, New York, 1993, pp. 377–396.

(24) Stach, E. *Lehrbuch der Kohlenpetrographie*, Borntraeger, Berlin, 1935

(25) Seyler, C. A.; Edwards, W. J. *Fuel*, **1949**, *28 (6)*, 121–126.

(26) Seyler, C. A. *Fuel*, **1952**, *31*, 133–152.

(27) Hoffmann, E.; Jenkner, A. *Gluckauf*, **1931**, *68*, 81–88.

(28) McCartney, J. T. *Econ. Geol.* **1952**, *47*, 202–210.

(29) Mukherjee, B. C. *Fuel*, **1952**, *31*, 153–158.

(30) Huntgens, F. J.; van Krevelen, D. W. *Fuel*, **1954**, *33*, 88–103.

(31) Siever, R. *Illinois Geol. Surv. Cir. 241*, **1957**.

(32) McCartney, J. T.; Teichmuller, M. *Fuel*, **1972**, *51*, 64–68.

(33) Berry, W. F.; Cameron, A. R.; Nandi, B. N. *Can. Inst. Min. Metall. Cent. Rev. Paper No. 1*, **1967**, 1–52.

(34) Davis, A. In *Analytical methods of coal coal products*, Kerr, C., Ed., Academic Press, New York, 1978, Vol. 1, pp. 27–81.

(35) Ting, F. T. C. In *Analytical methods for coal and coal products*, Kerr, C., Ed.; Academic Press, New York, 1978, Vol. 1, pp. 3–26.

(36) Teichmuller, M. In *Low temperature metamorphism*, Frey, M., Ed.; Chapman-Hall, Glasgow, 1987, pp. 115–160.

(37) Teichmuller, M.; Teichmuller, R. In *Diagenesis in Sediments and Sedimentary Rocks*, Larsen, G.; Chillingar, G. V., Eds., Developments in Sedimentology, Elsevier, Amsterdam, 1979, Vol. 25A, pp. 207–246.

(38) Teichmuller, M.; Teichmuller, R. *Bull. Cent. Rech, Explor, Prod. Elf Aquitaine*, **1981**, *5*, 491–534.

(39) Bustin, R. M.; Cameron, A. R.; Grieve, D. A.; Kalkreuth, W. D. *Coal Petrology, its Principles, Methods, and Applications*, Geological Association of Canada Short Course Note, 1985, Vol. 3.

(40) American Society for Testing and Materials (ASTM) In *Annual Book of Standards. 1991, Section 5, Vol. 5.05, Gaseous Fuel, Coal and Coke*, ASTM, Philadelphia.

(41) Houseknecht, D. W. *American Chemical Society 206 National Meeting, Chicago Abstract*, 1993, Abs. No. 27.

(42) Stout, S. A.; Spacman, W. *Int. Jour. Coal Geol.* **1987**, *8*, 55–68.

(43) Mukhopadhyay, P. K.; Hatcher, P. G. In *Hydrocarbons from Coal*, Law, B. E.; Rice, D. D., Eds.; Am. Assoc. Petrol. Geol. Studies in Geology Series, 1993, Vol. 38, pp. 79–118.

(44) Stanton, R. W.; and Moore, T. A. *TSOP Newsletter*, **1991**, *8 (1)*, 8–11.

(45) Mukhopadhyay, P. K.; Wade, J. A. *Bull. Can. Soc. Petrol. Geol.* **1990**, *38 (4)*, 407–425.

(46) Mukhopadhyay, P. K. *American Chemical Society 206 National Meeting, Chicago*, 1993, Abstract No. 6.

(47) Mukhopadhyay, P. K. *Rep. Invest. Texas Bur. Econ. Geol.* **1989**, *188*, 118p.

(48) Rimmer, S. M.; Davis, A. *Org. Geochem.* **1988**, *12*, 375–387.

(49) Hagemann, H. W.; Wolf, M. *Int. Jour. Coal Geol.* **1989**, *12*, 511–522.

(50) Stasiuk, L. D. *Organic Petrology and Petroleum Formation in Paleozoic rocks of Northern Williston Basin, Canada.* 1991, Ph. D. Thesis, University of Regina, Saskatchewan, Canada.

(51) Stasiuk, L. D.; Osadetz, K. G.; Goodarzi, F; Gentzis, T. *Int. Jour. Coal Geol.* **1991**, *19*, 457–481.

(52) Buchardt, B.; Lewan, M. D. *Bull. Am. Assoc. Petrol. Geol.* **1990**, *74 (4)*, 394–406.

(53) Barker, C. E. *TSOP Abstracts from the Annual Meeting, Jackson, Wyoming*, 1994.

(54) Hutton, A. C.; Cook, A. *Fuel*, **1980**, *59*, 711–716.

(55) Kalkreuth, W. D. *Bull. Can. Soc. Petrol. Geol.* **1982**, *30*, 112–139.

(56) Newman, J.; Newman, N. A. *New Zealand Jour. Petrol. Geol. Geophys.* **1982**, *25*, 233–243.

(57) Buiskool–Toxopeus, J. M. A. In *Petroleum Geochemical Exploration in Europe*, Brooks, J., Ed.; Blackwell, Oxford, 1983, pp. 295–307.

(58) Price, L. C.; Barker, C. E. *Jour. Petrol. Geol.* **1983**, *8 (1)*, 59–84.
(59) Goodarzi, F.; Nassichuk, W. W.; Snowdon, L. R.; Davies, G. R. *Mar. Petrol. Geol.* **1987**, *4*, 132–145.
(60) Kalkreuth, W. D.; Macaley, G. *Bull. Can. Petrol. Geol.* **1987**, *35 (3)*, 263–295.
(61) Wenger, L. M.; Barker, D. R. *Org. Geochem.* **1987**, *11*, 411–416.
(62) Goodarzi, F.; Gentzis, T.; Snowdon, L. R.; Bustin, R. M.; Feinstein, S.; Labonte, M. *Mar. Petrol. Geol.* **1993**, *10*, 162–171.
(63) Wilkens, R. W. T.; Wilmshurst, J. R.; Russel, N. J.; Hladky, G.; Ellkott, M. V.; Buckingham, C. *Org. Geochem.* **1992**, *18*, 629–640.
(64) Zhang, E; Hatcher, P. G.; Davis, A. *Org. Geochem.* **1993**, *20(6)*, 721–734.
(65) Lewan, M. D. *TSOP Abstracts from the Annual Meeting, Norman, Oklahoma*, 1993, 1–2.
(66) Lo, H. B. *Org. Geochem.* **1993**, *20 (6)*, 653–657.
(67) Mukhopadhyay, P. K.; Hagemann, H. W.; Gormly, J. R. *Erdol und Kohle–Erdgas–Petrochemie*, **1985**, *38(1)*, 7–18.
(68) Stout, S. A.; Boon, J. J.; Spackman, W. *Geochim. Cosmochim. Acta* , **1988**, *52*, 405–414.
(69) Larter, S. R. *Geol. Rundsch.*, **1989**, *78(1)*, 349–359.
(70) Senftle, J. T; Larter, S. R.; Bromley, B. W.; Brown, J. H. *Org. Geochem.* **1986**, *9(6)*, 345–350.
(71) Stout, S. A. *Int. Jour. Coal Geol.* **1993**, *24*, 309–331.
(72) Lin, R.; Ritz, G. P. *Org. Geochem.* **1993**, *20(6)*, 695–706.
(73) Stasiuk, L. D.; Kybett, B. D.; Bend, S. L. *Org. Geochem.* **1993**, *20(6)*, 707–719.
(74) Karweil, J. Z. *Dtsch. Geol. Ges.* **1956**, *107*, 132–139.
(75) Lopatin, N. V. *Akad. Nauk. S. S. S. R., Ser. Geol.* **1971**, *3*, 95–106.
(76) Tissot, B. *Rev. Inst. Franc. Petrol.* **1969**, *24*, 470–501.
(77) Waples, D. W. *Bull. Am. Assoc. petrol. Geol.* **1980**, *64*, 916–926.
(78) Connan, J. *Bull. Am. Assoc. Petrol. Geol.* **1974**, *58*, 2516–2521.
(79) Hood, A.; Gutjahr, C. M.; Peacock, R. L. *Bull. Am. Assoc. Petrol. Geol.* **1975**, *59*, 986–996.
(80) Bostick, N. H.; Cushman, S. M.; McCulloh, T. H.; Waddell, C. T. In *Low temperature metamorphism of kerogen and clay minerals*. Oltz, D. F., Ed.; 1978, SEPM Special Publication, pp. 65–96.
(81) Barker, C. E.; Pawlewicz, M. J. In *Paleogeothermics*, Buntebarth, G. and Stegena, L., Eds.; Springer–verlag, New York, 1986, pp 79–228.
(82) Barker, C. E. In *Thermal history of sedimentary basins – methods and case histories*, Naeser, N. D.; McCulloh, T. H., Eds.; Springer–Verlag, New York, 1989, pp. 75–98.
(82) Middleton, M. F. *Geophy. Jour. Royal Astrn. Soc.*, **1982**, *68*, 121–132.
(83) Welte, D.H.; Yalcin, M. N. *Org. Geochem.* **1988**, *13(1–3)*, 141–151.
(84) Welte, D. H.; Yukler, M. A. *Bull. Am. Assoc. Petrol. Geol.* **1981**, *65*, 589–617.
(85) Yukler, M. A.; Cornford, C.; Welte, D. H. *Geol. Rundsch.*, **1978**, *67*, 960–979.
(86) Yukler, M. A.; Kokesh, F. In *Advances in Petroleum Geochemistry*, Brooks, J.; Welte, D. H., Eds.; Academic Press, London, 1984, pp. 69–113.
(87) Lerche, I.; Yarzab, R. F.; Kendall, C. G. St. C. *Bull. Am. Assoc. Petrol. Geol.* **1984**, *68*, 1704–1717.

(88) Ritter, U. *Org. Geochem.* **1984**, *6*, 473–480.
(89) Armagnac, C.; Bucci, T.; Kendall, C. G. St. C.; Lerche, I. In *Thermal history of sedimentary basins – methods and case histories.* Naeser, N. D.; McCulloh, T. H., Eds.; Springer–Verlag, New York, 1989, pp. 217–238.
(90) Quigley, T. M.; Mackenzie, A. S.; Gray, J. R. In Migration of Hydrocarbons in Sedimentary Basins, Editor, B. Durand; Edition Technip, Paris; pp. 649–665.
(91) Larter, S. R. *Mar. Petrol. Geol.* **1988**, *5*, 194–204.
(92) Burnham, A. K.; Sweeney, J. J. *Geochim. Cosmo. Acta*, **1989**, *53*, 2649–2657.
(93) Sweeney, J. J.; Burnham, A. K. *Bull. Am. Assoc. Petrol. Geol.* **1990**, *74 (10)*, 1559–1570.
(94) Platt River Associates Inc. In *BasinMod™ – A Modelar Basin Modelling System, PC–DOS Version 3.15*, 1992, Boulder, Colorado.
(95) Suzuki, N.; Matsubayashi, H; Waples, D. W. *Bull. Am. Assoc. Petrol. Geol.* **1993**, *77 (9)*, 1502–1508.
(96) Staplin, F. L. *Soc. Econ. Paleont. Mineral. Short Course Note 7.*
(97) Heroux, Y.; Chagnon, A.; Bertrand, R. *Bull. Am. Assoc. Petrol. Geol.* **1979**, *63*, 2128–2144.
(98) Staplin, F. L. *Bull. Can. Jour. Pet. Geol.* **1969**, *17*, 47–66.
(99) Jones, R. W.; Edison, T. A. In *Low temperature metamorphism of kerogen and clay minerals.* Oltz, D. F., Ed.; 1978, SEPM Special Publication, pp. 1–12.
(100) Ottenjann, K. *Zeiss Inform.* **1981**, *26(93E)*, 40–46.
(101) Ottenjann, K. *Org. Geochem.* **1988**, *12(4)*, 309–322.
(102) Hagemann, H. W.; Hollerbach, A. *Bull. Cent. Rech. Explor. Prod. Elf–Aquit.* **1981**, *5*, 635–650.
(103) Teichmuller, M. *Geol. Lands. Nord. Westf. Spec. Publ.* Krefeld, Germany, 1982.
(104) Crelling, J. C. *Jour. Micros.* **1983**, *132(3)*, 135–144.
(105) Spiro, B.; Mukhopadhyay, P. K. *Erdol und Kohle–Erdgas–Petrochemie,* **1983**, *36(7)*, 297–299.
(106) Mukhopadhyay, P. K.; Rullkotter, J. *Bull. Am. Assoc. Pet. Geol.* **1986**, *70(5)*, 624.
(107) Thompson–Rizer, C. L.; Woods, R. A. *Int. Jour. Coal Geol.* **1987**, *7*, 85–104.
(108) Lin, R.; Davis, A. *Org. Geochem.* **1988**, *12(4)*, 363–374.
(109) Goodarzi, F.; Norford, B. C. *Int. Jour. Coal Geol.* **1989**, *11*, 127–141.
(110) Goodarzi, F. *Am. Chem. Soc. Nat. Meeting Chicago. Geochem. Divn. Abs.* 1993, 23.
(111) Goodarzi, F. *Mar. Petrol. Geol.* **1984**, *1*, 202–210.
(112) Goodarzi, F.; Norford, B. C. *Jour. Geol. Soc. Lond.* **1985**, *142*, 1089–1099.
(113) Goodarzi, F.; Higgins, A. C. *Mar. Petrol. Geol.* **1987**, *4*, 353–359.
(114) Goodarzi, F.; Snowdon, L. R.; Gunther, P.R.; Jenkins, W. A. *Mar. Petrol. Geol.* **1985**, *2*, 145–166.
(115) Bertrand, R.; Yeroux, Y. *Bull. A,. Assoc. Pet. Geol.* **1987**, *71(8)*, 951–957.
(116) Bertrand, R. *Am. Chem. Soc. Nat. Meeting Chicago Geochem. Divn. Abs,* 1993, 26.
(117) Jacob, H. *Erdol und Kohle–Erdgas–Petrochemie,* **1967**, *20*, 393–400.
(118) Jacob, H. *Int. Jour. Coal Geol.* **1989**, *11*, 65–79.
(119) Waples, D. W.; Machihara, T. *Biomarkers for Geologists, Am. Assoc. Petrol.*

Geol. Methods in Exploration No. 9, American Association of Petroleum Geologists Publication, Tulsa, Oklahoma, 1991.

(120) Radke, M. *Mar. Petrol. Geol.* **1988,** *5,* 224–236.

(121) Espitalie, J.; Deroo, Cr.; Marquis, F. *Rep. Inst. Frans. du Pertol.* **1985,** *33878,* 1–72.

(122) Radke, M.; Welte, D. H.; Willsch, H. In *Advances in Organic Geochemistry 1981;* Bjoroy, M. et al. Pergamon Press, Chichistor, England, 1983, pp. 504–512.

(123) Bein, A; Sofer, Z. *Bull. Am. Assoc. Petrol. Geol.* **1987,** *71,* 65–75.

(124) Goodarzi, F.; Brooks, P. W.; Embry, A. F. *Mar. Petrol. Geol.* 1989, *6,* 290–302.

RECEIVED August 12, 1994

PETROGRAPHY: STANDARDIZATION, APPLICATIONS,   AND   LIMITATIONS

Chapter 2

Need for Standardization of Vitrinite Reflectance Measurements

K. F. DeVanney[1] and R. W. Stanton[2]

[1]USX Corporation, UEC Coal and Coke Laboratories,
4000 Tech Center Drive, Monroeville, PA 15146
[2]U.S. Geological Survey, 956 National Center, Reston, VA 22092

Results of interlaboratory comparisons of vitrinite reflectance measurements indicate additional need for standardizing (1) the type of vitrinite to be measured, (2) the details of the measuring technique, and (3) the optical equipment used. Vitrinite reflectance is a critical characterization parameter that is widely used in geologic studies, exploration, classification, extraction, and utilization of fossil fuels. Precision can be a limiting factor in the use of vitrinite reflectance data, particularly when reflectances are measured in different laboratories. Whereas a value of 0.02 percent mean-maximum reflectance is the generally accepted level of reproducibility between two different laboratories, actual round-robin tests indicate that this level of precision may not be achievable even among those laboratories that follow a strict protocol. Results from the use of ASTM (American Society for Testing and Materials) interlaboratory training kits and test programs indicate that, at best, reproducibility is about 0.05 percent reflectance at best and can be as high as 0.15 percent reflectance.

Vitrinite reflectance is an important parameter for many areas of coal science and characterization. Because of the popularity of this measurement parameter, the conformance to standard methodology, equipment, and terminology are essential for comparisons to be made among reflectances measured by different workers. This paper will highlight some major applications for measuring vitrinite reflectance in coal and rock and the importance of standardization. Data from several round-robin test programs, the variables that affect vitrinite reflectance measurement, and recommendations to improve agreement between labs for measuring vitrinite reflectance for coal will also be discussed.

Uses of Vitrinite Reflectance

Coal classification, exploration, mining, beneficiation, and utilization, energy

0097–6156/94/0570–0026$08.00/0
© 1994 American Chemical Society

prospecting, and other geologic studies utilize, to various degrees, vitrinite reflectance and maceral composition. The use of reflectance data is of particular importance for commercial activities (coal buying and selling) especially in the coal, steel, and petroleum industries. Because reflectance data can be generated by various organizations, the values used must be comparable, technically equivalent, and independent of the analytical source. Comparability of data is critical because of the increased use of commercial laboratories by both producer and user of coal, and the need of industry to use various other sources of published data (i.e., government, universities) with confidence.

Coal Classification. Coal can be classified according to the American Society for Testing and Materials (ASTM) Classification of Coals by Rank D 388 (*1*) on the basis of mineral-matter-free (mmf) fixed-carbon contents (dry basis) and calorific value (moist basis). Rank can range from lignite to anthracite. In a recent revision of the ASTM standard (1), data are included that show the relationship between vitrinite reflectance and volatile matter on a dry, mmf basis (Fig. 1). This relationship shows a band of correlation that is about 0.2 percent reflectance wide. Rank classification is determined by bulk fixed-carbon content of a coal sample and therefore is a measure of the level of thermal maturity or degree of coalification of a heterogeneous mixture of macerals. In contrast, the reflectance of vitrinite is a measure of the degree of coalification of a single maceral. The width of the band in Fig 1. along the vertical axis (volatile matter), therefore, is indicative of the variation of the average composition of coals at a given level of thermal maturity.

Vitrinite reflectance is an important parameter in international commercial coal trade and scientific characterization of coal and rocks. A new codification system, replacing the previous international system adopted in the 1950's, has been developed by the United Nations' Economic Commission for Europe (ECE) (*2*). This codification system consists of 16 digits to characterize properties applicable to all coals for end-use utilization and commercial trade. A classification scheme for coals is being considered by the ECE; this classification is based on rank, type, and grade. In this classification and the codification schemes, vitrinite reflectance is a rank determinant for bituminous and higher rank coals.

Coal Exploration. Many companies conduct detailed drilling to provide information such as coal thickness, roof and floor conditions, seam depth, and quality assessment. Drill core analysis can provide coal rank and other key information in advance of property development (*3*). Vitrinite reflectance, in addition to other technical data, can provide information important to market suitability and mine planning.

Coal Mine Planning. Coal mine planning can utilize vitrinite reflectance and other data from both drill core and mine channel samples. One important use of the data is to produce reflectance isopleths to assist engineers in planning mine entries, mining direction, and mine development to (1) enable short- and long- term forecasting of production and quality and (2) maintain an average mine product quality through blending if the variation in reflectance is significant.

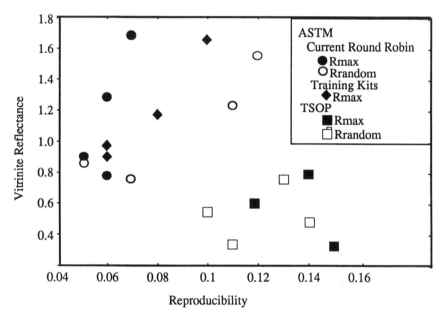

Figure 1. Plot of mean maximum reflectance of vitrinite and volatile matter on a dry, mineral-matter-free basis. [adapted from ASTM Classification of Coals By Rank (1); reflectance data are based on values from all types of vitrinite in each coal sample; hatched field represents the plots of 807 samples; Bit, bituminous.]

Coal Beneficiation. Samples of mined coal can be obtained from various preparation plant circuits for efficiency studies. Reflectance distributions of the coal, and particularly maceral concentrations, are sensitive to various circuits and process modifications because of the variation in densities among maceral groups. Many alterations can be made to adapt a product to meet specifications using coal petrographic data.

Coal Utilization - Cokemaking. On the basis of the applied research work by Schapiro and others (4) and Benedict and others (5), reflectance and maceral composition have been used to predict coking properties since the early 1960's. These same relationships are still in wide use today. Quality control in coke plants uses computerized systems to group and select coals on the basis of petrographic parameters (primarily reflectance and inertinite content) that are determined in advance of each shipment (6). Some coke plants also use automated reflectance data (displayed in a reflectogram--a histogram of reflectances), which are determined on coal blend samples prior to charging coal to coke ovens to confirm that the correct proportions have been blended. Both activities help in producing the required coke quality and in the prevention of coke oven damage.

Oil/Gas Prospecting. Geologists studying oil and gas and petrologists studying organic materials commonly rely on vitrinite reflectance data of both coal and dispersed organic matter to establish the level of thermal maturity for source rock units for hydrocarbon potential (7-10). This technique has been very useful in establishing windows for natural gas and petroleum occurrences in stratigraphic sequences.

Other Uses. Coal preparation, combustion, liquefaction, and gasification are other areas of application for vitrinite reflectance and petrographic data that will play a more important industrial role in the future (11-13) as conventional fuel resources become less abundant, or as refinements in these technologies require more specific characterization of the coals .

ASTM and ISO Standards for Measuring Vitrinite Reflectance

The two published standards that are commonly used and referenced for determining vitrinite reflectance are ASTM Test Method D 2798 (14) and the International Organization for Standardization (ISO) Document 7404-5 (15). For the most part, these two methods are technically equivalent. The more dynamic of the two documents is the ASTM Test Method, which is revised on a more frequent basis than the ISO Document. ASTM standards are required to be reapproved every 5 years; Test Method D 2798 was revised substantially in 1991 to reflect common practice. Table I shows a comparison between the ASTM and ISO maceral classification and terminology (15, 16). The ASTM Test Method defines the term vitrinite but does not subdivide the vitrinite group. The ISO document subdivides the vitrinite maceral group into three vitrinite macerals, which in turn, are subdivided into 6 submacerals; however, the ISO terms are not defined in the ISO method. Definitions for those terms shown in the ISO document are found in the International Committee for Coal Petrology (ICCP) Glossary (*17*). Although the protocols for methods are under the direct control of ISO process, coal petrographic terminology is under the control of ICCP. As a result, the ICCP terms do not receive the same type of scrutiny as other terms receive in ISO documents. Other standardized terminology and modified classification schemes also exist, and these are almost as numerous as the users. Terminology for the vitrinite maceral group alone is quite variable, which provides additional reasons for the need for international agreement on specific terms used in vitrinite reflectance.

Interlaboratory Comparison of Data

Standards provide a benchmark by which researchers can directly compare measurements, provided that the protocols specified in the standard are followed precisely. Table II shows summaries of various interlaboratory reflectance comparison programs to illustrate the magnitude of differences among labs.

Table I.   **Maceral Terminology**
        [ASTM, American Society for Testing and Materials; ISO, International Organization for Standards]

ASTM (D-121)		ISO (7404-1)	
Maceral Group	Maceral	Maceral	Submaceral
Vitrinite	Vitrinite	Telinite	Telinite 1
			Telinite 2
		Collinite	Telocollinite
			Gelocollinite
			Desmocollinite
			Corpocollinite
		Vitrodetrinite	
Liptinite	Alginite	Alginite	
	Cutinite	Cutinite	
	Resinite	Resinite	
	Sporinite	Sporinite	
		Suberinite	
		Liptodetrinite	
		Bituminite	
Inertinite	Fusinite	Fusinite	Pyrofusinite
			Degradofusinite
	Inertodetrinite		Inertodetrinite
	Macrinite	Macrinite	
	Micrinite	Micrinite	
	Sclerotinite	Sclerotinite	
	Semifusinite	Semifusinite	

Table II. Vitrinite Reflectance data for interlaboratory tests conducted by The Society for Organic Petrology (TSOP) and American Society for Testing and Materials (ASTM) Committee D5.
[s.d. = standard deviation; N=number of labs reporting data; diff=difference between the maximum and minimum values reported; Rep = reproducibility calculated as 2.8 times the standard deviation; values other than N are reported in percent]

	Mean	s.d.	N	Range	Diff	Rep
TSOP Round Robin (18)						
Mean Maximum Reflectance						
Lignite	0.32	0.054	9	0.27-0.39	0.12	0.15
High Volatile B Bituminous	0.57	0.039	9	0.49-0.62	0.13	0.11
High Volatile A Bituminous	0.79	0.050	9	0.71-0.87	0.16	0.14
Mean Random Reflectance						
Lignite	0.33	0.038	13	0.25-0.37	0.12	0.11
High Volatile B Bituminous	0.55	0.036	13	0.47-0.60	0.13	0.10
High Volatile A Bituminous	0.76	0.046	13	0.68-0.84	0.16	0.13
Organic-rich shale	0.47	0.050	13	0.27-0.39	0.12	0.14
ASTM Training Kit Program (19)						
Mean Maximum Reflectance						
High Volatile Bit. 14675	0.90	0.021	30	0.85-0.95	0.10	0.06
High Volatile Bit. 14527	0.97	0.023	30	0.91-1.01	0.10	0.06
Med. Volatile Bit. 14526	1.18	0.030	30	1.11-1.24	0.13	0.08
Low Volatile Bit. 14607	1.66	0.035	30	1.57-1.72	0.15	0.10
ASTM D05.28 Round Robin (20)						
(in progress)						
Mean Maximum Reflectance						
High Volatile Bit. 16364	0.78	0.023	15	0.73-0.82	0.09	0.06
High Volatile Bit. 16160	0.91	0.019	13	0.89-0.96	0.07	0.05
Med. Volatile Bit. 16349	1.28	0.021	14	1.24-1.32	0.08	0.06
Low Volatile Bit. 16183	1.68	0.025	12	1.63-1.72	0.09	0.07
Mean Random Reflectance						
High Volatile Bit. 16364	0.75	0.026	16	0.71-0.81	0.10	0.07
High Volatile Bit. 16160	0.85	0.017	13	0.83-0.88	0.05	0.05
Med. Volatile Bit. 16349	1.22	0.028	14	1.18-1.29	0.11	0.08
Low Volatile Bit. 16183	1.56	0.042	12	1.49-1.66	0.17	0.12

TSOP. A round-robin project was conducted by the Society for Organic Petrology (TSOP) Research Subcommittee to standardize reflectance and fluorescence methods. A total of 16 laboratories participated to test reproducibility and interlaboratory comparability of results within the TSOP membership (18). For reflectance measurements, each participant was asked to measure (1) the mean maximum reflectance of vitrinite for each of three coal samples by rotating the stage and using a polarizer, and (2) the mean random reflectance of vitrinite for each of four samples without rotating the stage or using a polarizer. No standard method was specified nor mentioned in the instructions.

On the basis of the worst case (lignite-Mean Max. R_o), the reproducibility calculated from the standard deviation was about 0.15 percent (table II). Therefore, comparing results between two labs at a 95 percent confidence level, the data are considered suspect if the difference exceeds 0.15 percent for a 0.32 percent reflectance coal. Similar differences are seen for the Mean Random reflectance data for lignite.

ASTM Training Kit Data Summary. As a part of protocol development in ASTM, a training kit was developed which contained six coal samples of different ranks. Participants were instructed to follow the 1979 version of Test Method D 2798 (*14*). Data were collected from 30 labs from 6 countries (Table II; *19*). The differences among labs increase with rank, but appear slightly better than the TSOP data. The improvement over TSOP reproducibility may be the result of less morphological variation in the vitrinite types in higher rank coals used in the ASTM Training Kit or closer adherence to a common procedure. On the basis of the worst case (low volatile bituminous - Mean Max. R_o), the reproducibility calculated from the standard deviation is about 0.10 percent. Therefore, comparing results between two labs at a 95 percent confidence level, the data are considered suspect if the difference exceeds 0.10 percent for a coal having a reflectance of 1.66 percent.

ASTM Round Robin. An ASTM task group is currently conducting a reflectance round robin to determine the precision to update the 1989 revision of Test Method D 2798 (*14*). Table II shows a data summary of the ASTM round robin (*20*). The mean maximum R_o results are summarized based on 11-15 labs from the USA and Canada. The differences among labs for each coal in all cases are less than the TSOP and ASTM Training Kit data and reproducibilities are better (Figure 2). This improved reproducibility is probably the result of fewer morphological complexities for the higher rank coals used, operator experience, and stricter adherence to ASTM standards being followed by these labs. For the worst case, reproducibility calculated from the standard deviation is about 0.07 percent for a low volatile bituminous coal, on the basis of available data. Larger differences can be observed in the mean random data as compared to mean maximum data. The last two samples in the ASTM interlaboratory comparison will be completed in 1994, bringing the total in this set to six samples. After statistical analysis and removal of technical outliers, precision statements will be published in the ASTM method in accordance with ASTM Practice E 691 (*21*) for within-lab repeatability and between-lab reproducibility.

It should be noted that the reproducibility data from these three different round-robin exercises range from about 0.05 percent to 0.15 percent. Considering that most

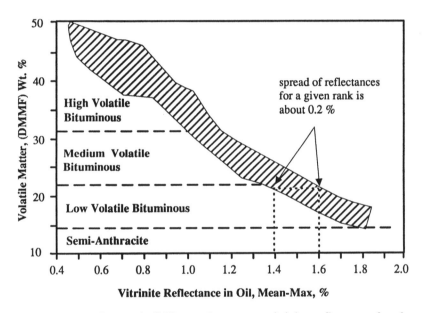

Figure 2. Plot of reproducibility against mean vitrinite reflectance for three interlaboratory sets of data from Table II.

labs estimate their repeatability (within lab) to be about 0.02 percent, some of the reproducibility data generated are unacceptably high.

Variables Affecting Reflectance Data

The differences among labs for vitrinite reflectance need to be reduced. Most of the sources of variation among various organizations fit into the following broad categories: sample preparation, microscope equipment and optical alignment, type and use of calibration standards, and the selection of vitrinite type to be measured.

Sample Preparation Variables. Many of the differences in reflectance measurements that can be attributed to sample preparation result from differences in the following: coal particle size used in the briquette, polishing technique used (time, pressure, polishing media and lap cloths), briquette pressurization during impregnation with epoxy, epoxy hardness, contamination, desiccation of the coal, and general skill of the technician. One very important area of standard revision for both ASTM and ISO methods is in specific procedure steps to increase the quality of the sample briquettes and surface polish used for reflectance analysis.

Microscope Equipment and Optical Alignment. In both the ASTM and ISO methods, the use of microscope equipment is described. Use of the correct equipment, proper alignment, and strict adherence to the published standards (ASTM and ISO) are critical. Most of the equipment specified is obtainable from microscope suppliers. Of critical importance is that the lamp used must be powered by a stabilized source and that the vertical illuminator must be similar to either a Berek prism or Smith glass type. Plane-glass vertical illuminators cause light scatter and difficulty in measuring low-reflecting macerals. Additionally, the correct interference filter must be used, and no field or aperture diaphragms should be used during reflectance measurements.

Calibration Standards. The actual technique of calibration with reference materials must be further refined. When calibrating for measuring mean-maximum reflectance, some workers calibrate by rotating the microscope stage just as is done during an actual measurement, whereas others calibrate without any rotation of the stage. The use of more than one calibration standard to bracket the range being measured should be common practice. Recommended procedures for the care and calibration of glass standards have been added to ASTM Test Method D 2798-91 as an appendix (nonmandatory information) (14).

Dependable sources for the supply of glass reflectance standards must be found. Because the refractive indices even on standard glasses differ both within a glass and among different glasses, particularly on a microscopic level, verification of the refractive indices for calibration standards supplied should accompany the standard. The refractive indices for calibration standards should be supplied as a certificate of analysis by a standards organization or other credible entity that houses primary reflectance standards. Both ASTM and ISO Test Methods should be revised to be more specific in this area.

Vitrinite Selection and Type. Terms used for specific vitrinite types or submacerals measured for reflectance can be inconsistent and confusing among labs. For example, vitrinite A and B, telinite, collinite, gelinite, pseudovitrinite, telocollinite, desmocollinite, gelocollinite, vitrodetrinite, and the huminite series are terms commonly used by various labs. ISO Method 7404/5 (*15*) states that the reflectances of the submacerals of vitrinite differ even in a single coal seam and therefore it is necessary to specify on which submacerals the measurements were made.

Vitrinite is composed of coalified cell walls, humic cell fillings, and humic matrix gels; these various types have different refractive indices when observed in thin section. Hence various types will also have different reflectances. Many varieties of vitrinite can exhibit similar appearance in reflected light, but in transmitted light, seemingly homogeneous vitrinite may show a more complex vitrinitic substance having varying refractive properties (Figure 3). In another example, vitrinite bands can be observed to be mottled in some parts and seemingly homogeneous in other layers; etching reveals that the bands are composed of different vitrinite types. In Figure 4, mottled vitrinite can be observed to consist of humic gels and cell fillings, whereas other layers consist of compressed cell walls. Vitrinite submacerals can be difficult to identify and measure because of rank, variety, and maceral size. Etched surfaces can assist in vitrinite submaceral identification when this type of differentiation is required (*22*). However, identification of vitrinite types prior to etching can be difficult in all cases. Some labs measure all types of vitrinite (ASTM) whereas others may measure only certain types. ASTM Test Method D 2798 states that the particular vitrinite type measured for reflectance shall be specified when used as a rank indicator, but these vitrinite types are not currently defined in ASTM.

To date, there are no universally accepted terms or methods for measuring reflectance for coals below bituminous rank, particularly lignites. These coals present even more problems for vitrinite submaceral classification. In lignite, additional problems arise from the presence of porosity and degrees of gelification of the humic material. For these coals, a separate maceral classification standard and reflectance test method may be necessary.

Conclusions

The use of vitrinite reflectance is growing in importance internationally. Most of the differences in reflectance measurements between labs are from sample preparation, microscopic equipment and alignment, calibration and standards, and selection of vitrinite type. These variables are controllable and can be standardized. Because of the variety of uses for vitrinite reflectance data and the international use of this data as a measure the degree of coalification, the needs for standardization and improvement are highly important now. To achieve better reproducibility, the following steps are necessary:

1. Differences among labs in the reproducibilities of mean maximum and random reflectance measurements of vitrinite must be decreased.

2. Current standard methods must continue to be improved and terminology expanded and made uniform.

3. Current standards must be implemented and strictly followed in labs that measure vitrinite reflectance.

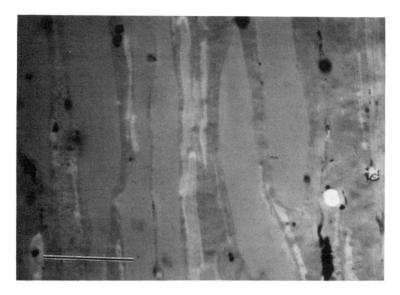

Figure 3. Photomicrograph of thin section of vitrinite types of a high volatile bituminous coal from the Appalachian Basin. [In color, most of these components are various shades of red. Scale bar equals 50 μm.]

Figure 4. Reflected light photomicrograph of polished vitrinite which has been etched by acidified potassium permanganate (on right part of photomicrograph). [Variation in etching response is indicative of minor differences in chemical composition. Scale bar equals 50 μm].

4. Participation in standards organizations (for example ASTM and ISO through ASTM) must increase.
5. Standardization should be expanded to include lower rank coals, organic-matter-rich rocks and sediments, and new techniques such as etching.

Literature Cited

(1)   American Society for Testing and Materials (ASTM), *Standard Classification of Coals by Rank, Designation: D388-92A, Annual Book of ASTM Standards 1993*, Vol. 05.05, pp. 199-202.

(2)   United Nations (Geneva), Economic Commission for Europe, *International Codification System for Medium and High Rank Coals,* Document ECE/COAL/115, United Nations: New York, 1988.

(3)   *Manual on Drilling, sampling, and analysis of coal,* Luppens, J.A., Wilson, S.E., and Stanton, R.W., Eds., ASTM Manual Series MNL11, ASTM: Philadelphia, PA, 1992.

(4)   Schapiro, N., Gray, R.J., and Eusner, G.R., *Blast Furnace, Coke Oven and Raw Materials Committee, Proceedings,* 1961, Vol. 20, pp. 89-109.

(5)   Benedict, L.G., Thompson, R.R., and Wenger, R.O., *Blast Furnace and Steel Plant,* **1968**, vol 56, pp. 217-224.

(6)   Graeser, W.C., McCollum, H.R., and Huntington, H.D., *AIME Ironmaking Conference Proceedings.* 1989, Vol. 48, pp. 97-102.

(7)   Teichmuller, M. and Teichmuller, R. In *Diagenesis in Sediments*; Larsen, G. and Chilingar, G.V., Eds.; Elsevier: Amsterdam, 1967, pp. 391-415.

(8)   Teichmuller, M., and Teichmuller, R. In *Coal and Coal-bearing Strata*; Murchison, D.G., and Westoll, T.S., Eds.; Oliver and Boyd: Edinburgh, 1968; pp. 233-267.

(9)   Tissot, B.P. and Welte, D.H., *Petroleum Formation and Occurrence*; Springer-Verlag: Berlin, 1978.

(10) Robert, P, In *Kerogen: Insoluble Organic Matter from Sedimentary Rocks*; Durand, B., Ed.; Editions Technip: Paris, 1980; pp. 385-414.

(11) Vleeskens, J.M., *Effect of Coal Properties on the Fluidized Bed Combustion Efficiency of Single Coals and Blends,* Netherlands Research Foundation, ECN-133, May 1983.

(12) Murchison, D.G., In *Coal and Coal-bearing Strata: Recent Advances*; Scott, A.C., Ed.; Geological Society Special Publication No. 32, pp. 257-302.

(13) Khorasani, G.K., In *Coal and Coal-bearing Strata: Recent Advances*; Scott, A.C., Ed.; Geological Society Special Publication No. 32, pp. 303-310.

(14) American Society for Testing and Materials (ASTM), Standard Test Method for Microscopical Determination of the Reflectance of Vitrinite in a Polished Specimen of Coal, Designation: D2798-91, *Annual Book of ASTM Standards* 1993, Vol. 05.05, pp. 309-312.

(15) International Organization for Standardization (ISO) Methods for the petrographic analysis of bituminous coal and anthracite- Part 1: Glossary of terms, Ref. No. ISO 7404/1-1984 (E) and Part 5: Method of determining microscopically the reflectance of vitrinite, Ref No. IS) 7404/5-1984 (E).

(16) American Society for Testing and Materials (ASTM), Standard Terminology of

Coal and Coke, Designation: D 121-92, *Annual Book of ASTM Standards 1993*, Vol. 05.05, pp. 169-179.

(17) International Committee on Coal Petrology (ICCP) *International Handbook of Coal Petrography*, 2nd Ed., Centre National de la recherche Scientifique Academy of the USSR, Paris, Moscow, 1963.

(18) Lin, R., *The Society for Organic Petrology Newsletter*, 1993, *vol 9*, No. 4, .

(19) ASTM D05.28 Subcommittee on Coal and Coke Petrography, unpublished data on training kits 1984-1993; permission to use these preliminary findings was given by Lou Janke, Chairman D-5 Committee on Coal and Coke.

(20) ASTM D05.28 Subcommittee on Coal and Coke Petrography, unpublished data on interlaboratory Study 1992-1993; permission to use these preliminary findings was given by Lou Janke, Chairman D-5 Committee on Coal and Coke.

(21) American Society for Testing and Materials (ASTM), Standard Practice Conducting an Interlaboratory Study to determine the Precision of a Test Method, Designation: D691-87, *Annual Book of ASTM Standards 1993*, Vol. 14.02, pp. 430-449.

(22) Stanton, R.W., and Moore, T.A., *The Society for Organic Petrology Newsletter*, 1991, *vol 8*, no. 1, pp. 8-11

RECEIVED July 28, 1994

Chapter 3

Presence of Remanent Cell Structure in Vitrinite and Its Influence on Reflectance Properties

D. F. Bensley and J. C. Crelling

Department of Geology, Southern Illinois University, Carbondale, IL 62901

The interpretation of reflectance is dependent in part upon an assessment of the reflectance distribution. In general, as the distribution of reflectance values broaden, displaying an increase in standard deviation, the analytical interpretation becomes increasingly difficult. Data scatter would traditionally be attributed to recycled vitrinite, oxidized vitrinite, multiple vitrinite populations, misidentification, analytical problems, vitrinite reflectance suppression or similar sources of error. It is of interest, however, that a broad reflectance distribution is seldom attributed solely to the heterogeneity of the vitrinite macerals themselves. It is concluded that part of the scatter in reflectance values obtained for a given coal is due to the presence of remanent cell structure. This structure can not only be observed through etching but can be directly measured using reflectance. Sequential spatially orientated reflectance readings using rotational polarization reflectance has been used to document cell structures unobservable with conventional microscopic analysis.

The limitations inherent in analyzing samples processed using density gradient centrifugation, DGC, necessitated the development of new quantitative tools to permit accurate characterization of ultra-fine coal particles. Correlations between reflectance distributions obtained using random reflectance on both -20 mesh vitrinite and corresponding vitrinite separations from the same sample were considered marginal at best. In some samples the relationship between random reflectance on DGC vitrinite separations and the vitrinite samples from which separations were made could not be clearly ascertained. Because reflectance distributions using

0097–6156/94/0570–0039$08.00/0
© 1994 American Chemical Society

maximum reflectance tend to show less variation than random reflectance, it was reasoned that a more accurate reflectance value could be obtained if maximum reflectance was measured.

Equipment and analytical techniques were developed to measure maximum reflectance on particles less than $2\mu m$. During testing, multiple reflectance measurements on single, apparently homogeneous vitrinite macerals were observed to produce highly variable results. Reflectance variability, as expected, increased with decreasing measuring point size. This variability, however, could not be attributed solely to analytical error or equipment response. One possible explanation for the observed variability in measured reflectance is the presence of underlying remanent cell structure. Several studies have shown that much of the actual morphology present in vitrinite is obscured during polishing, (1,2,3,4,5,6). If this hidden cell structure possesses distinctive reflectance properties, then measurements made on vitrinite would vary depending upon the unobserved heterogeneity of the maceral.

Experiments were carried out to determine reflectance variability in single vitrinite macerals and any possible relationship to underlying structure. Sequential measurements were made across apparently homogeneous telocollinite with both the maximum reflectance and the spatial orientation of the point stored within a data file. After the initial reflectance measurement, the pellet was then etched. An oxidizing solution was prepared as described in Stach, et al, (7) and Moore and Stanton, (8). For this experiment 25 grams of $KMnO_4$ dissolved in 70 mL of H_2O and mixed with 5 grams H_2SO_4 were prepared. Pellet etching was carried out by immersing polished pellets into the prepared solution. To facilitate etching the solution was heated to 50° C. Etching times were found to be rank dependent and varied from 40 minutes for high-volatile C bituminous coals to 5 hours for low-volatile bituminous samples.

Equipment

Reflectance data was obtained using a Leitz MPVIII compact microscope modified for rotational polarization reflectance, (9); (Bensley and Crelling, *Fuel.*, in press). The principle alteration from a typical reflectance microscope is the adaptation of the polarizer to permit rotation through 360 degrees. All polarizer rotation, data acquisition and data processing functions are computer controlled using a 80386/20 mHz computer. Microscope interfacing is made via a gear coupled polarizer and stepping stage, stepping stage controller and IEEE bus. A separate A/D converter controls data acquisition. Optical correction is made on all measurements to remove the effect of residual polarization within the vertical illuminator and transmission optics.

Results and Discussion

Figures 1 and 2 illustrate the results obtained from sequential reflectance measurements. The area delineated with grids represent point locations where individual reflectance measurements were made. The reflectance distribution and average maximum reflectance values are given for each telocollinite maceral. It should be noted that in each example there is at least a 1 V-type range in reflectance values. Figure 1 illustrates the distribution of maximum reflectance readings from 346 sequential measurements. An overall average reflectance of 0.99% was obtained for this maceral with individual reflectance values ranging from 0.95% to 1.06%.

In figure 2 two adjacent telocollinite macerals were evaluated. Both macerals possess similar overall reflectance properties with an average reflectance of 0.97%, (256 readings), and 0.94%, (366 readings), noted for each maceral. Both macerals illustrated in figure 2 possess a distinctive bimodal reflectance distribution which contrasts from the distribution observed in the telocollinite maceral illustrated in figure 1. In all three cases telocollinite macerals from the same coal, possessing similar average reflectance, 0.99%, 0.97% and 0.94% respectively, show marked differences in the distribution of reflectance values observed for each maceral.

Figure 3 illustrates the effect of etching on the maceral analyzed in figure 1. Comparison between the etched structure and the reflectance values obtained on the polished maceral are illustrated in figure 4. Figure 4 uses gray scale imaging to facilitate comparison between reflectance values and morphology in the maceral itself. In figure 4 a reflectance range has been assigned to a gray scale value and positioned to the appropriate spacial orientation. Comparison of the white light images in figure 3 and the coded, mapped reflectance area of figure 4 demonstrate some measure of influence on reflectance of the remanent cell structure revealed through etching. The low reflectance values, noted by dark values, correspond with areas etched away by the oxidizing solution. This is particularly noticeable in the right section of the image and the center where low reflectance values tend to correspond to open cell structure. The highest reflectance values also correspond to the areas of higher reflectance preserved in the etched sample. Similar correlation of reflectance values to maceral structure in the unetched image is not evident.

Figure 5 illustrates the effects of etching on the two telocollinite macerals shown in figure 2. In this case there are marked differences between the properties of the two macerals despite similar average reflectance. The right maceral is extensively etched by the solution while the maceral on the left is preserved very much intact. In figure 6 the reflectance distributions are shown against the gray scale image of the unetched and etched macerals shown in figure 5. As noted previously, areas of extreme etching correspond to regions of lower reflectance in the

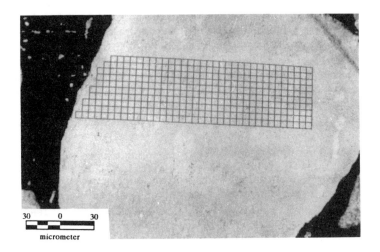

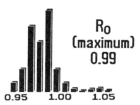

Figure 1: Reflectance distribution as measured on a apparently homogeneous telocollinite.

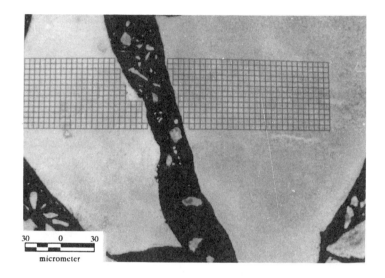

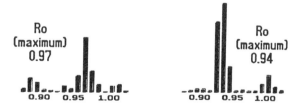

Figure 2:   Reflectance distributions measured on two adjacent telocollinite macerals.

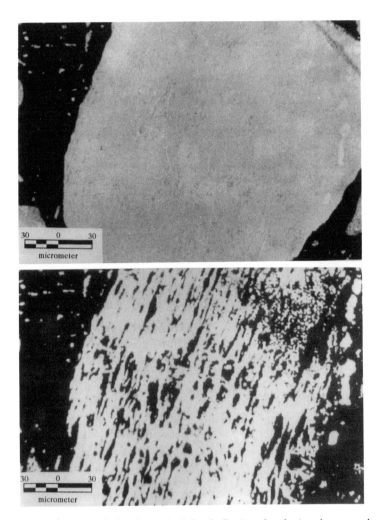

Figure 3: Unetched, (top), and etched, (bottom), photomicrograph of a telocollinite maceral illustrated in figure 1.

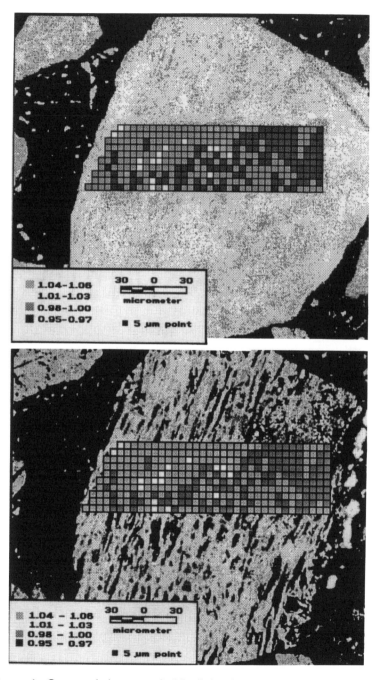

Figure 4: Gray scale images of white light photomicrographs illustrated in figure 3 superimposed with coded reflectance map illustrated in figure 1.

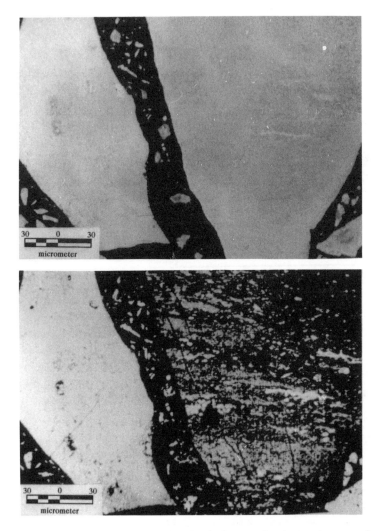

Figure 5: Unetched, (top), and etched, (bottom), photomicrograph of the two telocollinite macerals illustrated in figure 2.

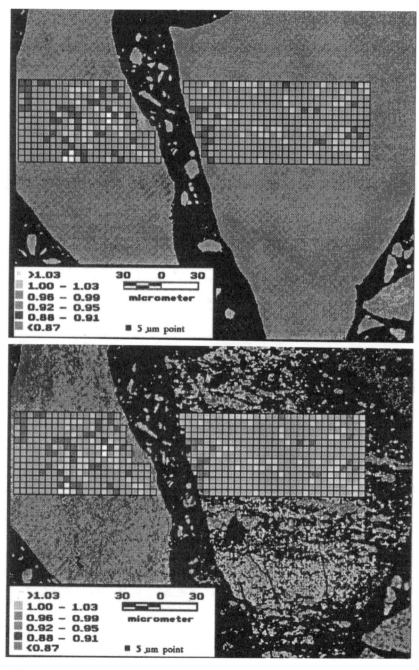

Figure 6: Gray scale images of white light photomicrographs illustrated in figure 5 superimposed with coded reflectance maps illustrated in figure 2.

maceral shown in the left field of the figure. Areas of higher reflectance correspond to the most resistant regions in the etched maceral. Perhaps the most striking feature of the image is that areas possessing reflectance of intermediate value correspond very well to regions undergoing minor alteration after etching. This is seen most easily in the left most area of the image where an arc of intermediate reflectance values runs between lower reflecting regions.

The maceral positioned to the right of the image illustrated in figure 5 lacks the clear associations with remanent cell structure to reflectance values noted in the left maceral. The lower reflecting areas in the left most field of the right image do correspond to regions of extensive etching. Most of the maceral, however, has been severely etched and little cell structure remains evident.   Correlation between either the unetched telocollinite or the etched structure is difficult to ascertain.

Figure 7 shows the effect of variation in the measuring spot size on vitrinite reflectance. It is evident that as the area of measurement increases, the resultant reflectance distribution tightens. The observed increase in standard deviation noted with decreasing measuring spot size has generally been attributed to signal degradation. The interpretation from reflectance mapping suggests that heterogeneity within the vitrinite can account for much of the observed scatter in the reflectance distribution.   Figure 8 illustrates the effect of the measuring point size. During stage rotation the measuring point will migrate across an area which can represent hidden morphology of variable reflectance.   The resulting reflectance value represents an average of the area measured with extreme high or low values effectively averaged out. Similarly, when larger measuring areas are used, reflectance tends to converge on an intermediate value.   The precision of the measurement is therefore controlled by the size of the measuring point and the ellipse of rotation or measuring point size. When a small measuring point is used, the odds of landing on a single remnant cell is increased, hence extreme high or low values are more readily observed within the reflectance distribution. This interpretation is consistent with the data collected and can readily explain the observed increase in standard deviation noted for random reflectance, which typically uses small measuring points, when compared with conventional maximum reflectance measurement.

Conclusions

The variation in vitrinite reflectance observed in single macerals can approximate the variation in reflectance for the whole coal. This variability appears to be caused by remnant cell structure within the vitrinite maceral itself. This heterogeneity is not readily observed since polishing obscures the morphology. Etching the polished sample is the only practical way this hidden morphology can be visualized.

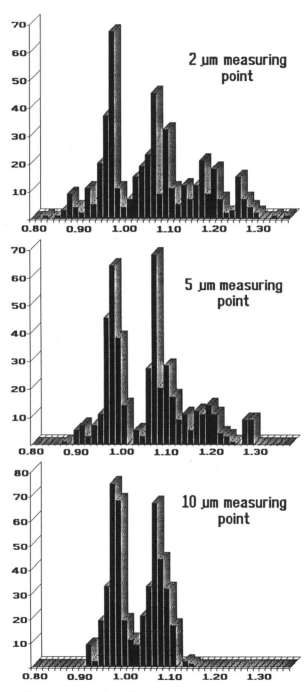

Figure 7:   The variation in reflectance distribution noted for various
measuring point sizes. All readings were taken from the same vitrinite
maceral.

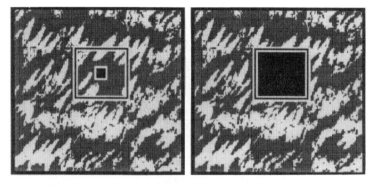

Figure 8: Illustration of measuring point size as it relates to remnant cell structure.

The size of the measuring point controls the reflectance distribution and the effect of remnant cell structure on the distribution. As the measuring point size increases, the macerals inherent heterogeneity converges on an average. The larger the area of measurement the tighter the observed reflectance distribution.

Not all vitrinite macerals show structural influence but many do. The precise influence of remnant structure on a typical reflectance analysis is unknown at this time. Data suggests that an increase in the reflectance distribution around the reflectance average should occur. The mean reflectance should not, however, change significantly.

Literature Cited

1. Stach, E. and Kuhlwein, F. L.; Die mikroskopische Untersuchung feinkörniger Kohlenaufbereitungsprodukte in Kohlenreliefschliff. - Glückauf 64, 1928, 841-845.
2. Stach, E.; Lehrbuch der Kohlenmikroskopie. - Verlag Glückauf: Kettwig, 1949, 285 pp.
3. Mackowsky, M.-TH.; Das Ätzen der Steinkohle, ein aufschlußreiches Hilfsmittel in der Kohlenmikroskopie. - C. r. 7. Congr. intern. Strat. Géol. Carbonifère 3, Krefeld, 1974, 375-383.
4. Hacquebard, P. A., Birmingham, T. F. and Donaldson, J. R.; Petrography of Canadian coals in relation to environment of deposition. - Symp. Science and Technol. of Coal, 1967, 87-97, Ottawa.
5. Pierce, B. S., Stanton, R. W., and Eble, C. F.; Facies development in the Lower Freeport coal bead, west-central Pennsylvania, U.S.A. - Int. J. Coal Geol. 1991, 18, 17-43.
6. Pierce, B. S., Stanton, R. W., and Eble, C. F.; Comparison of the petrography, palynology and paleobotany of the Stockton coal bed, West Virginia and implications for paleoenvironmental interpretations. - Org. Geochem. 1993, 20, 2. 149-166.
7. Stach, E., Mackowsky, M.TH., Teichmüller, M., Taylor, G. H., Chandra, D., Teichmüller, R.; Stach's Textbook of Coal Petrology. - Gebrüder Borntraeger, 1982, 535 pp.
8. Moore, T. A., Stanton, R. W.; Coal petrographic laboratory procedures and safety manual. - U.S. Geological Survey Open-File Report 1985, 85-20.
9. Bensley, D. F. and Crelling, J. C.; The use of rotational polarization reflectance in the characterization of fine particulate macerals: Proceedings of the Int. Conf. on Coal Science, 1993, v. 1., 578-581.

RECEIVED April 15, 1994

Chapter 4

Corpohuminite from Canadian Paleozoic Source Rocks

Petrology and Reflectance Properties

L. D. Stasiuk, F. Goodarzi, and M. G. Fowler

Geological Survey of Canada, 3303 Thirty-Third Street Northwest, Calgary, Alberta T2L 2A7, Canada

Corpohuminite macerals from four Paleozoic hydrocarbon source rocks of Canada were studied using reflected light microscopy. Perennial algal akinete cells produced by cyanophytes during algal bloom episodes are proposed as the precursor organic matter for these macerals. Per cent reflectance in oil (%Ro) of the corpohuminites (0.24-0.90 %Ro) is similar to the reflectance of associated vitrinite (0.21-0.86 %Ro) in Devonian and Mississippian rocks. The reflectance increases with increasing burial depth and increasing thermal maturity parameters (e.g. Tmax from Rock-Eval pyrolysis). Over the range of corpohuminite reflectance Tmax ranges from 412 to 445°C. For sedimentary rocks barren of vitrinite, the algal akinetes may be used as an accessory for assessing the thermal maturity and for constraining thermal maturity levels in terms of vitrinite reflectance and hydrocarbon generation.

The lack of true vitrinite derived from the lignin of terrestrial plants is a frequent problem when optically (per cent reflectance in oil- %Ro) assessing the thermal maturity levels of carbonate, marine or essentially any pre-Devonian potential hydrocarbon source rocks and associated sedimentary strata. Earlier studies have reported on the reflectance in oil of vitrinite-like macerals in lower Paleozoic (1, 2). Due to evolution constraints, these macerals are clearly not derived from the lignin of vascular plants. The reliability and reproducibility of using the reflectance of vitrinite-like macerals as a universal thermal maturation indicator still remains to be proven.

The purpose of this study is to summarize the petrology and reflectance properties of algal-derived corpohuminite-like bodies (hereafter referred to as corpohuminite) from Paleozoic potential hydrocarbon source rocks from three

0097–6156/94/0570–0052$08.00/0
Published 1994 American Chemical Society

sedimentary basins in Canada (Figure 1). Corpohuminite macerals in coal are defined as cell infillings, with huminite/vitrinite reflectivities, liberated as excretions from plant cell walls or poduced as secondary cell infillings of humic gels[3]; algal-derived tannins are also a source of corpohuminite[4]. The corpohuminite macerals in the Canadian Paleozoic rocks provide: (i) information on paleoenvironmental conditions (i.e. algal blooms) and; (ii) constraint on the thermal maturity level with respect to petroleum generation. The formations of study include the Upper Ordovician-Silurian Cape Phillips Formation, Franklinian Basin (Truro Island, Arctic), Middle Devonian Winnipegosis and Devonian-Mississippian Bakken Formation (Williston Basin, Saskatchewan) and the Middle Devonian Keg River Formation (La Crete Basin, Alberta) (Figure 1). A summary of the source rock potential of the formations is given in Table I.

Table I.   Summary of Average Source Rock Characteristics from Rock-Eval pyrolysis

Formation	Lithology	T.O.C.[a]	Hydrogen index[b]	Oxygen index[c]
Cape Phillips	black shale	5.57	488	6
Key River	carbonate laminite	14.50	485	26
Winnipegosis[d]	carbonate laminate	7.38	515	33
Bakken[d]	black shale	14.70	404	30

[a]Total organic carbon; [b]mg hydrocarbons/g org. C x 100; [c]mg CO_2/g org. C x 100; [d](*19*).

Experimental

Oriented (parallel and perpendicular to bedding) and crushed core samples (2-4 mm diameter) were mounted in epoxy resin, ground and polished using a modified version of the procedure described by Mackowsky (*5*). A Zeiss MPM II incident light microscope (visible and ultraviolet light sources), a Zeiss 03 photomultiplier unit, Zonax computer, Zeiss Continuous Filter Monochromator b and an epiplan neofluor oil immersion objective (x40; N.A. 0.90 oil) were used for data collection. Random reflectance in oil (mean %Ro; n = 1.5138) of macerals was measured on the particulate samples. The number of reflectance measurements ranged from 2-50 and standard deviations from 0.02-0.09 (*6*).

Figure 1. Location map of the four Paleozoic formations included in this study: A- Williston Basin (Winnipegosis and Bakken Formations), Saskatchewan; B- La Crete Basin (Keg River Formation), Alberta; C- Franklinian Basin, Truro Island (Cape Phillips Formation).

Results and Discussion

Petrography. The morphological characteristics of corpohuminite macerals from the Middle Devonian Winnipegosis Formation and the Devonian-Mississippian Bakken Formation of the Williston Basin have been described in detail by Stasiuk (*7*). Identical macerals were found in two other Paleozoic potential hydrocarbon source rock intervals, the Middle Devonian Keg River Formation of the La Crete Basin, Alberta, and the Upper Ordovician-Silurian Cape Phillips Formation, Truro Island Arctic, Canada. Corpohuminites have also been noted in source rocks of the Middle to Upper Ordovician Collingwood Formation, Ontario, Canada (*8*).

The corpohuminites are typically cigar-shaped, ovoid or rounded, are rarely vacuolated and range in size from 2 to 12 μm in diameter. They most commonly occur as isolated entities within a network of amorphous, bituminite-dominated kerogen (Figure 2a) but occasionally they form part of clusters and chains comprising several corpohuminite bodies (Figure 2b). Abundant unicellular Prasinophyte alginite (*Tasmanites* and *Leiosphaeridia*) and agglomerations of micrinitic bacteria-sized coccoidal cells (~ 1 μm in diameter) are commonly associated with the corpohuminites (Figure 2c).

Highly specialized, non-vegetative cells (perennial) derived from cyanobacteria ("blue-green algae") have been proposed as the precursors of corpohuminite macerals in Paleozoic rocks (*6, 7*). These cells, referred to as algal akinete cells, are formed by extant cyanobacteria during periods of environmental stress, most notably during periods of algal blooms (*9, 10*). The presence of corpohuminites in Paleozoic rocks may be strong evidence for paleo-algal blooms. Macerals associated with the corpohuminites and the preservational habit of these macerals is also supportive of the algal bloom theory (e.g. abundant unicellular Prasinophyte and bacteria; *6, 7*). For example populations of degraded unicellular alginite in the source rocks are interpreted as products of mechanical and chemical degradation of planktonic organisms through cell lysis (bacterial and/or photochemical alteration), an hyperactive phenomena during blooms (*6, 11*). The planktonic degradation (i.e. of Prasinophyte alginites) results from a consequential bacteria population explosion (*6, 11*) and abnormal "overfloating" of the plankton into the upper portion of the water column in search of energy for photosynthesis; in this zone the plankton are severely photo-oxidized (*6*). Present day cyanobacteria respond to such counter-growth conditions by producing akinete cells, which, because of increased density, sink to the bottom, only to germinate once conditions become normal.

Reflectance properties. The reflectance data for corpohuminites from a sample with a good population number typically has a normal distribution much like the distribution for a primary population of dispersed vitrinite macerals (*12*). Figure 3 illustrates the range and frequency distribution of corpohuminite reflectance measurements for an organic-rich sample of the Middle Devonian Keg River Formation from Alberta. Reflectance of the corpohuminite macerals versus depth for the Winnipegosis and Bakken Formations of the Williston Basin (Figure 4a,

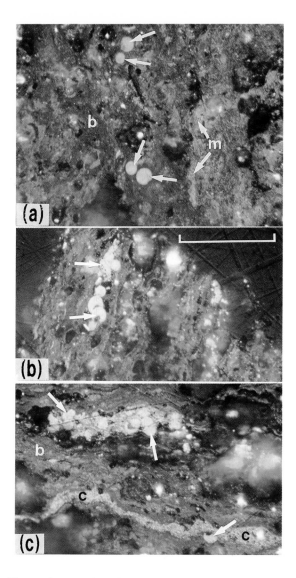

Figure 2. Photomicrographs of corpohuminites (as indicated by arrows) proposed to be derived from algal akinete cells of cyanobacteria. The corpohominites are associated with bituminite III (b), micrinite (m) and laminations of bacteria-like coccoidal cells (c); oil immersion, plane polarized light, scale bar = 50 μm.

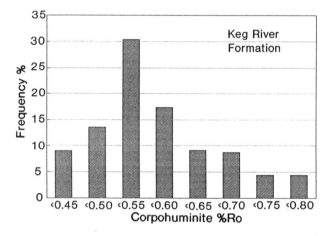

Figure 3. Reflectance histogram for %Ro of corpohuminites in Middle Devonian Keg River Formation, Alberta; n = 41, mean = 0.55, std. dev. = 0.09.

b), display a pattern similar to the vitrinite reflectance profiles but with a much less pronounced %Ro increase with depth (*13*). A positive correlation between vitrinite reflectance and corpohuminite reflectance for the two formations corroborates the sensitivity of the algal akinete cells to variations in thermal maturation (Figure 5a, b). A positive correlation also exists between the range of mean corpohuminite reflectance (0.26 to 0.90 %Ro) and the range in Tmax values from Rock Eval pyrolysis (413 to 445 °C) (Figure 6a, b), supporting the proposal that algal akinete cell reflectance can be used for constraining the thermal maturity of hydrocarbon source rocks.

Distinct low and high slope vitrinite reflectance gradients have been defined for the Winnipegosis and Bakken Formations (*13*); subtle high and low gradients are suggested by similar plots of the reflectance of corpohuminite versus depth (Figure 4a, b). The distribution of corpohuminite reflectance for the Winnipegosis and Bakken Formation in the Williston Basin also resembles the regional maturation pattern as defined by vitrinite reflectance. The similarity in the regional thermal maturation pattern as defined by vitrinite (*13*) and corpohuminites is particularly striking for the Bakken Formation where the higher %Ro values occur in the southern central portion of the study area and the lower %Ro values occur in the western to northern portion of the study area (Figure 7a, b).

The positive correlation between vitrinite and algal akinete cell reflectance may result from the fact that the maceral precursors had similar chemical composition. Akinete cells of extant cyanobacteria are formed through the condensation of the mucilaginous sheath and the deposition of a very dense fibrillar layer consisting of polyglucosides and a large amount of cyanophycin and glycogen granules (*14*). The link between the chemical composition of akinetes as a precursor to a vitrinite maceral needs further investigation; some algae are known to produce phenolic moieties (tannins) under specific environmental conditions, which like lignin are characterized by aromatic (phenolic) structures (*15*).

Corpohuminite-like macerals (as *tanins*) are included in Alpern's (*16*) classification for organoclasts testifying to their importance as dispersed macerals in sedimentary rocks other than coal. The reflectance of corpohuminites has been shown to be similar to other vitrinite macerals at least from coals of lignite to sub-bituminous rank (*17, 18*). It is therefore no surprise that the corpohuminite macerals from the Paleozoic would also accommodate thermally maturity assessment. The correlations illustrated here indicate that the reflectance of algal akinete cells provides a constraint for selecting primary populations of vitrinite for reflectance, and with further work, may potentially be a reliable independent indicator of thermal maturity levels.

Summary

Corpohuminite macerals from organic-rich Paleozoic rocks of Canada are interpreted to be derived from algal akinete cells of cyanobacteria ("blue-green" algae). These highly specialized perennial cells are produced by cyanophytes

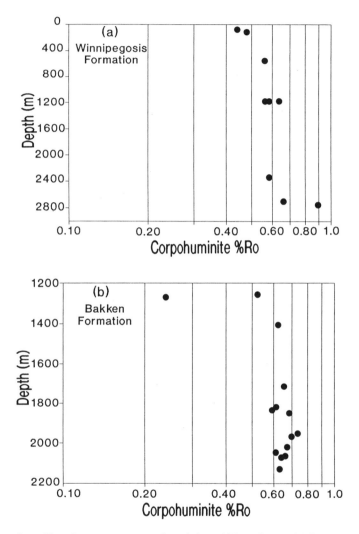

Figure 4.   Depth versus corpohuminite %Ro for Middle Devonian Winnipegosis (a) and Devonian-Mississippian Bakken (b) Formations, Williston Basin.

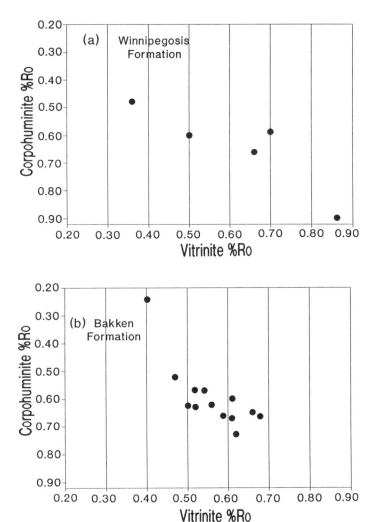

Figure 5.   Vitrinite %Ro versus corpohuminite %Ro in oil for the
Winnipegosis (a) and Devonian-Mississippian Bakken (b) Formations,
Williston Basin.

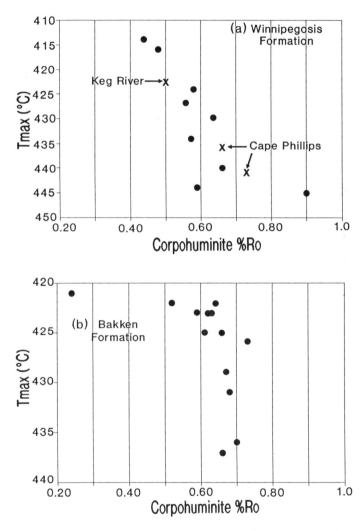

Figure 6. Corpohuminite %Ro versus Tmax (°C) from Rock Eval pyrolysis for the Winnipegosis (a) and Bakken (b) Formations, Williston Basin; data for the Cape Phillips Formation, Franklinian Basin and Keg River Formation, La Crete Basin are also included in (b).

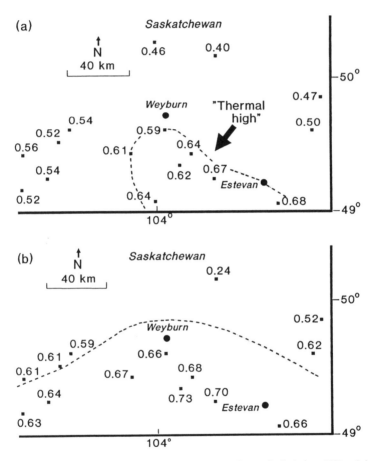

Figure 7. Comparison of the regional distribution of vitrinite %Ro (a) and corpohuminite %Ro (b) for the Bakken Formation at sample locations (■) in southeastern Saskatchewan.

during periods of abnormal environmental conditions such as algal blooms. The reflectance in oil of the corpohuminites (range: 0.24 to 0.90 %Ro) increases with depth and thermal maturity and is similar to that of associated vitrinite. The reflectance of the corpohuminites is sensitive for determining regional variations in levels of thermal maturation. An important implication of this study is that for sedimentary rocks barren of vitrinite, the reflectance of algal akinetes may be used as an accessory parameter for assessing the thermal maturity. This maceral is still under study and further detailed geochemical and reflectance work is anticipated.

References

1. Sikander A.H.; Pittion, J.L. *Bull. Can. Pet. Geol.* 1978; *26*, pp. 131-151.
2. Buchardt, B.; Lewan, M.D. *Bull. Amer. Assoc. Petrol. Geol.* 1990; *74*, pp. 394-406.
3. Mackowsky, M.-Th. In *Stachs textbook of coal petrology*; Stach, E. *et al.*. Eds.; Gebruder Borntraeger: Berlin, 1982; 2nd Edition; pp. 295-299.
4. Teichmüller, M. In *Stachs textbook of coal petrology*; Stach, E. *et al.*, Eds.; Gebruder Borntraeger: Berlin, 1982; 2nd Edition; pp. 176-228.
5. International Commission of Coal Petrology. *International Handbook of Coal Petrology*, 2nd Edition; Centre National de la Recherche Scientifique, Paris.
6. Stasiuk, L.D. Ph.D. Thesis, University of Regina, 1992.
7. Stasiuk, L.D. *Int. J. Coal Geol.* 1993; *24*, pp. 195-210.
8. Mark Obermajer. University of Western Ontario, personal communication, 1993.
9. Fay, P. *The blue-greens*; Institute of Biology's Studies in Biology 160; Camelot Press: Southampton, 1983; 88 p.
10. South, G.R.; Whittick, A. *Introduction to phycology*; Blackwell Scientific Publications: Oxford, 1987; 341 p.
11. Round, F.E. *The ecology of algae*; Cambridge University Press: Cambridge, 1981; 653 p.
12. Mukhopadhyay, P.K. In *Diagenesis III*; Wolf, K.H.; Chilingarian, G.V., Eds.; Developments in sedimentology 47, Elsevier, 1992; pp. 435-510.
13. Stasiuk L.D. (*Bull. Can. Petrol. Geol.* in press).
14. Nicols J.M.; Adams D.G. In *Biology of Cyanobacteria*; Whitton B.A.; Carr N.G., Eds.; Blackwell Scientific Publications: 1982; pp. 389-412.
15. Tissot B.P.; Welte D.H. Petroleum formation and occurrence;. Springer-Verlag: Berlin, 1984; 699 p.
16. Alpern, B. *Rev. L'Institut Franc. Pétrol.* 1970; *25*, pp. 1233-1266.
17. Goodarzi, F.J. *Petrol. Geol.* 1987; *10*, pp. 219-226.
18. Goodarzi, F.; Gentzis, T.; Feinstein S.; Snowdon, L. *Int. J. Coal Geol.* 1988; *10*, pp. 383-398.
19. Osadetz K.G.; Brooks P.W.; Snowdon L.R. *Bull. Can. Petrol. Geol.* 1992; *40*, pp. 254-273.

RECEIVED June 6, 1994

Chapter 5

Iso-rank Variation of Vitrinite Reflectance and Fluorescence Intensity

J. C. Quick

Earth Sciences and Resources Institute, University of South Carolina,
901 Sumpter Street, Room 401, Columbia, SC 29208

Between 0.5 and 1.1% mean maximum reflectance of vitrinite ($\overline{Ro}_{max}$), the log mean fluorescence intensity of vitrinite is shown to exhibit a negative linear relationship with iso-rank variation of vitrinite reflectance. This relationship is used to devise a method to both identify reflectance-suppressed vitrinite and to estimate the extent of suppression. Iso-rank variation of vitrinite reflectance and fluorescence intensity observed in one Late Cretaceous and three Eocene age coal seams is attributed to the activity of methanogenic and sulfate-reducing bacteria.

The popular use of vitrinite reflectance to estimate the maturity of source rocks can be attributed to several factors: vitrinite exhibits a wide temporal and spatial distribution, the measurement is relatively unaffected by associated mineral, kerogen, or maceral assemblages, and it is sensitive to organic maturation through the oil window. Although accurate reflectance determinations are routinely accomplished, vitrinite reflectance, like all maturity parameters, is sometimes misleading.

Reflectance analyses undertaken on kerogen preparations that show multiple reflectance populations may indicate the presence of reworked vitrinite, drillhole caving contamination, or native bitumen. Recognition of multiple vitrinite populations may enable better interpretation of other geochemical analyses but where the objective is to estimate maturity, the analyst must subjectively identify the indigenous population. Where indigenous vitrinite populations are readily identified other problems may hinder accurate maturity estimation. Between 0.4 and 0.8% $\overline{Ro}_{max}$, pore-hosted moisture lowers reflectance (1) and may impart a patchy, mottled appearance (2). Where patchy vitrinite is encountered, avoiding measurements on the darker patches minimizes the problem; simple desiccation of the specimen is not always effective (unpublished data of author). Reflectance will also vary according to the relative abundance of different vitrinite group macerals (3); the reflectance of pseudovitrinite, oxyvitrinite, telocollininte and phlobaphinite, is typically greater than associated desmocollinite, detrovitrinite or degradovitrinite. This variation is reduced where reflectance is deter-

0097–6156/94/0570–0064$08.00/0
© 1994 American Chemical Society

mined on a single maceral (e.g., telocollinite) but this option is not always possible. Reflectance variation within morphologically similar vitrinite macerals is less obvious. For example, Goodarzi and others (4) observed reflectance variation according to the lithology of the enclosing sediment rather than the morphology of the vitrinite maceral.

Although anomalous reflectance can sometimes be attributed to variation within the vitrinite group or the lithology of the enclosing sediment, objective criteria that quantify extent of reflectance variation are less common (5-6). In this paper the fluorescence intensity of vitrinite is used to both identify and correct for the extent of reflectance suppression.

Development of Fluorometric Analysis

Fluorometric analysis is undertaken using an incident light microscope and reports the mean fluorescence intensity of a selected maceral population. Visibly fluorescent vitrinite (7-9) was first observed in studies using BG (blue-glass) excitation filters rather than shorter wavelength UG (long-wave UV) glass filters used for spectral fluorescence studies. Dichroic beamsplitters, which direct the excitation light to the specimen and more than triple the fluorescence emission compared to 50/50 half silvered reflectors, were also found to be important. Early quantitative measures of fluorescence intensity (10-11) were accomplished by calibrating with an uranyl glass standard. The use of longer wavelength (450-490nm) interference type excitation filters (12), allowed fluorescence intensity measurements to be undertaken on both vitrinite and inertinite group macerals. High performance interference filters offer three significant advantages compared to colored glass filters, including: 1) high transmission efficiency in the bandpass region, 2) the availability of custom designed filters (Omega Optical Co. Brattleboro VT), and 3) significantly improved blocking outside the bandpass region. Improved blocking is important since even small amounts of unwanted long-wave light that leaks through the excitation filter will overwhelm weak fluorescence signals. Gross leakage is indicated where sulfide grains appear faint red in the fluorescence observation mode.

A more recent advance in fluorometric technique is the elimination of oxygen from the specimen surface during fluorometric analysis. The use of a nitrogen or argon flow over the specimen surface eliminates negative alteration (fading) upon irradiation of the surface with blue excitation light (13). This phenomenon requires that fluorescence intensity measurements be obtained immediately upon exposure of the specimen surface to the excitation energy to obtain repeatable results. Because 50 or more measurements may be used to calculate the mean vitrinite fluorescence intensity, frequent switching between white and blue light illumination modes was required. An inert gas flow over the specimen circumvents the need for frequent switching between white and blue light and reduces both measurement error and analysis time.

Method

Samples. Core samples were obtained from four coal seams occurring in the west coast region of South Island, New Zealand (Figure 1). The seam occurring in drillhole (DH) 712 in the Greymouth coalfield is of Late Cretaceous age; the remaining seams are Eocene age. DH-1481 is from a quality monitoring program from the actively producing no.2 block in the Buller Coalfield; DH-7 is an exploration drillhole from the remote Pike River coalfield, and DH-1494 is an exploration drillhole in the Upper Waimangaroa sector of the Buller Coalfield.

Each coal seam was divided into vertically adjacent increments to obtain a set of samples corresponding to the entire thickness of the seam. Analyses undertaken on the samples comprising each set show the vertical variation of a property through that seam. Because all the samples in a set share the same burial history they are called iso-rank samples.

Like many Pacific Tertiary age coals, the samples in DH-7, DH-1494, and DH-1481 show high vitrinite contents and little inertinite (14). Coal samples from DH-712 (Late Cretaceous) contain slightly less vitrinite (avg. 91%) and higher inertinite (avg. 6%). The vitric nature of these coals minimizes analytical variation of bulk chemical analyses due to group maceral variation. Consequently bulk chemical analyses undertaken on these samples can be used to approximate vitrinite composition. Nonetheless, even small variations in liptinite content (0-5%) have been correlated with hydrogen abundance in these coals (*15*). Thus, whole coal or bulk chemical analyses of vitrinite-rich samples may approximate vitrinite composition but the precision of such analyses is limited.

Analytical Methods. Mean maximum vitrinite reflectance was determined in general accordance with AS-2486 (*16*). Proximate analyses and determination of sulfur forms were undertaken by a commercial laboratory.

Fluorometric analyses were undertaken using a Zeiss research microscope. A Wild Leitz masked uranyl glass standard was used for fluorescence calibration and arbitrarily assigned an intensity of 100. Where appropriate, individually calibrated coated neutral density filters were used during calibration to reduce the fluorescence intensity of the uranyl glass standard to levels comparable with the specimen and the data mathematically corrected to the uranyl glass =100 scale. This arrangement allowed fluorescence intensity as low as 0.01 to be measured; the visual threshold of fluorescence is near 1.5. No background correction was needed since attempts to measure fluorescence intensity of sulfide minerals present in the specimens gave values less than the detection limit (0.01 where calibrated for maximum sensitivity). Spectral characteristics of the excitation filter, dichroic mirror, and measurement filters are shown in Figure 2. The Zeiss USMP research microscope used in the study included a X40, 0.85 n.a. dry objective, a 100 W Hg arc lamp with stabilized DC power supply, and a R928 side-on photomultiplier tube operated with a nominal gain of 0.7 kV and a 0.16 second time constant. The measuring spot diameter was 9.4 μm and the field diaphragm was 15 μm. No heat filter was used. The measurement filter was placed below the oculars and also functioned as a barrier filter. This placement required the analyses to be undertaken in a darkened room but has the advantage of increasing the signal reaching the detector by about 20%. Raw data were electronically recorded with four digits of precision on a personal computer with custom data acquisition hardware (1.5 second integration time) and custom software. During analysis, a nitrogen flow of 1.7 liters N_2 per minute was delivered to the specimen surface through a custom manifold. Further details are provided elsewhere (*15*).

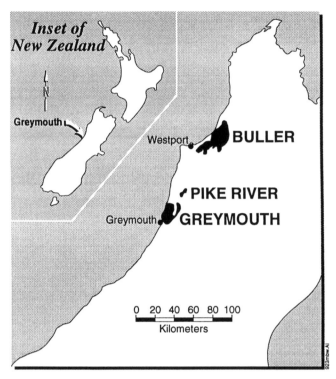

Figure 1. Map showing the location of coalfields the samples came from.

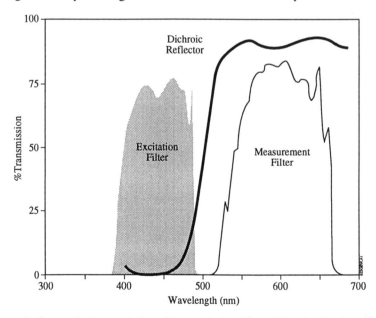

Figure 2. Spectral characteristics of the excitation filter, dichroic illuminator, and measurement filter used for fluorometric analysis.

Results

Figure 3 shows the vertical variation of vitrinite reflectance, fluorescence intensity and sulfur content through the four coal seams examined. Because the adjacent samples in a given drillhole shared the same burial history they should report the same vitrinite reflectance. Examination of Figure 3 shows that the vertical variation of mean vitrinite reflectance through a single seam can vary by as much as 0.20% which is 10 times greater than the nominal precision of the method (about 0.02%, AS-2486). Unlike the Eocene age coal seams, DH-712 shows a minimum reflectance in the seam center rather than seam margins.

Within each of the four coal seams shown in Figure 3, where vitrinite reflectance is low, fluorescence intensity is high. Vertical variation of sulfur through the three Eocene age coals generally corresponds to the variation of fluorescence intensity whereas the sulfur content of DH-712 is uniformly low and shows no correlation with either vitrinite reflectance or fluorescence intensity.

Figure 4 shows a cross plot of vitrinite reflectance and the log of fluorescence intensity for the sample set. Linear regression lines for each iso-rank suite of samples are also shown and report r^2 values of 0.84, 0.99, 0.89, and 0.96 for samples from DH-712, DH-1494, DH-7 and DH-1481, respectively. In addition, these lines share a similar slope. The nominal slope of the four lines shown in Figure 4 is -3.4.

Correction of Suppressed Vitrinite Reflectance. The relationship shown in Figure 4 can be used to estimate the extent of reflectance suppression in a given sample where both reflectance and log fluorescence intensity are known. However, before this can be done the relationship between vitrinite reflectance and fluorescence intensity needs to be defined. The fluorescence intensity of vitrinite waxes and wanes as maturation progresses through the bituminous ranks. This increase and subsequent decrease of vitrinite fluorescence intensity has been attributed to the build-up and subsequent destruction of a molecular mobile phase (17). Figure 5 shows the variation of vitrinite fluorescence intensity for a set of whole-seam US coals. Fluorescence intensity is shown to increase and decrease with increasing reflectance as well as to exhibit a wide natural variation in the high-volatile bituminous ranks (about 0.5 to 1.1% $\overline{R}o_{max}$).

Examination of Figure 5 shows a mathematically fit curvilinear line that follows the general increase and decrease of vitrinite fluorescence intensity with increasing reflectance. The position of this line would have been different had the data used to construct Figure 5 been obtained from different coals. For example, Diessel and Wolff-Fischer (18) show different reflectance/fluorescence distributions for vitrinite in Permian and Carboniferous age coals. Thus, the line showing the relationship between fluorescence intensity and vitrinite reflectance in Figure 5 is provincial. For this reason a constant vitrinite fluorescence intensity of 10 (log FLI=1) is used to quantify the extent of reflectance suppression. Although this values is arbitrary, it does approximate the average fluorescence intensity many coals of different ages and ranks examined by the author. It also avoids the equally arbitrary distinction of what provincial set of coals should be considered to be "normal".

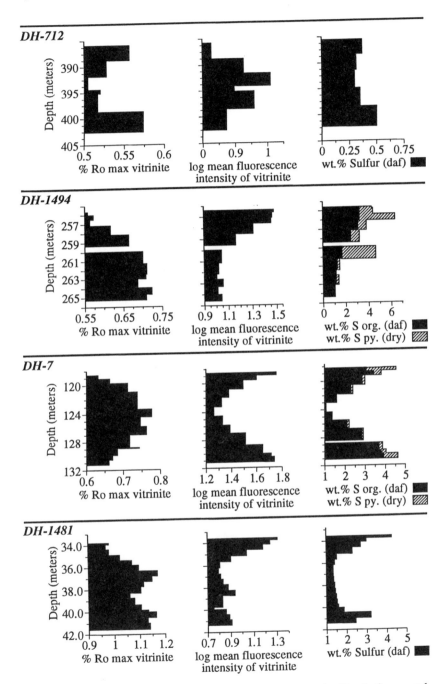

Figure 3. Vertical variation of reflectance, fluorescence and sulfur in four cored coal seams.

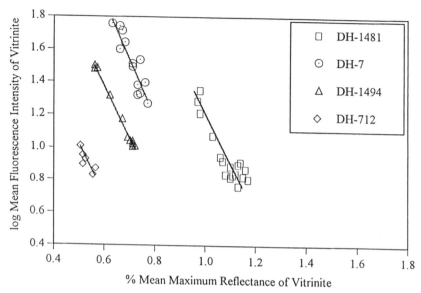

Figure 4. Vitrinite reflectance vs vitrinite fluorescence intensity for four cored coal seams.

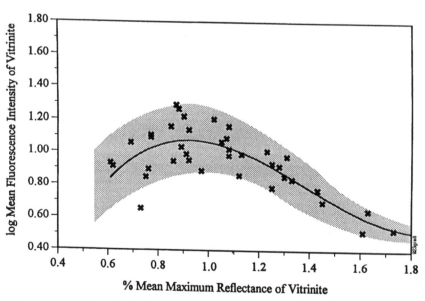

Figure 5. Variation of vitrinite fluorescence intensity for 38 whole seam US coals. Figure constructed from modified data presented in Ref. (19).

Using the average slope of the regression lines shown in Figure 4 and a constant base value of 10 for vitrinite fluorescence intensity the extent of reflectance suppression can be estimated as:

$$\text{Ro suppression} = (\log \text{FLI} - 1)/3.4 \tag{1}$$

where FLI is the mean fluorescence intensity of the vitrinite population and Ro suppression is the absolute percent of reflectance suppression for that sample relative to an equivalent sample with a mean fluorescence intensity of 10. If the result of equation 1 is added to the measured reflectance the result indicates what the reflectance would be at a constant fluorescence intensity. This correction is graphically illustrated in Figure 6 where a coal with measured reflectance of 0.56% $\overline{Ro}_{max}$ and a measured FLI of 31 (log FLI=1.5) is plotted. The corrected reflectance is obtained by graphically shifting this plotted point down a line with a slope of -3.4 to a position where the log fluorescence intensity equals 1. The adjusted reflectance is indicated on the x axis directly below this position.

The effectiveness of the proposed correction method is illustrated in Figure 7 where the measured reflectance profile through a coal seam encountered in DH-7 is compared to the reflectance profile of the same seam after correction. Examination of the figure readily shows that iso-rank variation of the measured reflectance is reduced upon correction to a level commensurate with the precision of reflectance analysis ($\pm$ 0.02).

Discussion

There are several limitations to the correction method. First, the method is only appropriate for vitrinite in samples with a mean reflectance between 0.5 and 1.1% $\overline{Ro}_{max}$. Although it may be possible to extend the relationship to include higher rank coals the lower limit is probably about 0.4% $\overline{Ro}_{max}$; low rank coals (lignite and brown coal) exhibit high fluorescence intensity unrelated to fluorescence in higher rank coals (*17,19*). A second limitation is that the method requires fresh samples; fluorescence intensity is diminished due to weathering (19), laboratory storage (*20-21*), stockpile storage (*15*), and even shallow present day burial depth (*22*). Thirdly, it is important to note that fluorometric analysis is not yet a standardized procedure; different instruments in different laboratories using different calibration standards give different numerical results but nonetheless show the same trends for the same samples (*22*). Finally, note that the procedure has not yet been applied to studies on dispersed organic matter where suppressed vitrinite reflectance is of concern. Dispersed organic matter is usually concentrated using acid demineralization and gravity separation techniques prior to microscopic examination. The effect of this treatment on vitrinite fluorescence intensity is not known.

Despite these limitations the method has several advantages. The correction can be applied to any sample where both vitrinite reflectance and vitrinite fluorescence intensity are known. Furthermore, it is entirely mathematical and does not require any graphical manipulation of data. In addition, because both reflectance and fluorescence

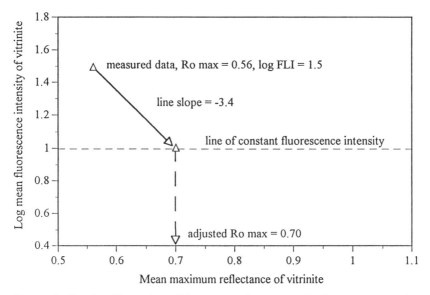

Figure 6. Graphic illustration of the proposed suppressed reflectance correction where a sample with a vitrinite reflectance of 0.56 and a fluorescence intensity of 31 (logFLI=1.5) is shown to have an adjusted reflectance of 0.70.

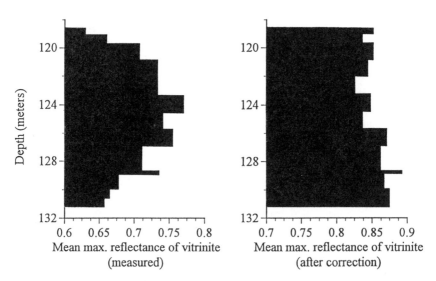

Figure 7. Comparison of the measured (left) and corrected (right) reflectance profiles for a seam encountered in DH-7.

intensity are measured on the vitrinite population, perturbations arising from the presence of liptinite or inertinite group macerals are avoided.

Origins of Perhydrous Vitrinite. The genesis of perhydrous vitrinite that exhibits suppressed vitrinite reflectance and high fluorescence intensity remains to be fully examined. Two different early biogenic alteration mechanisms have been suggested (*23*) and are discussed below.

The acidic conditions that characterize a coal-forming mire inhibit biological degradation of the peat. However, base exchange with adjacent inorganic sediments locally increases the pH at seam margins (*24*) and enables primary fermentive anaerobes to collectively decompose polymers to CO_2, H_2 and water. These bacteria thrive only where the product hydrogen is removed (*25*). Two kinds of bacteria compete for the hydrogen (*26*). Where sulfate is not available, methanogens convert the hydrogen to methane. Where sulfate is available, sulfate-reducing bacteria successfully compete for the hydrogen and hydrogen sulfide, rather than methane, is produced.

Methanogens. Methanogenic bacterial degradation results in a net loss of hydrogen from the coal-forming peat as methane is lost to the atmosphere. This mechanism can explain the seam profiles shown for DH-712. The overall low sulfur content for the seam indicates that sulfate ion was not abundant thus allowing for methanogenic bacterial growth. Base exchange from adjacent sediments (*24*) would elevate the pH at seam margins and allow growth of fermentive anaerobes (*27*), production of H_2, and its elimination as gaseous methane. Bacterial growth in the center of the seam was inhibited due to the acidic pore waters and inherited hydrogen was preserved. Thus preservation of the original hydrogen may explain the occurrence of low sulfur perhydrous coal. The influence of methanogenic bacteria on subsequent hydrogen preservation or depletion is likely to be more important in geologically young coals. This is because the relative amounts of cellulose and lignin, which are the major precursors of vitrinite, have changed over geologic time. Robinson (*28*) suggested that plants have become more sparing in their use of lignin while increasing cellulose production over geologic time. Since cellulose is hydrogen-rich compared to the more refractory lignin, the potential for variation of hydrogen content of vitrinite due to the activity of methanogenic bacteria is greater in geologically young coals. Given the susceptibility of cellulose to hydrolysis by fungi, this mechanism necessarily requires limited aerobic decay.

Sulfate Reducers. Where sulfate ions are present, sulfate reducing bacteria successfully compete with methanogenic bacteria for free hydrogen produced by the fermentive anaerobes. Sulfate-reducing bacteria convert hydrogen into hydrogen sulfide that, in the absence of reactive iron, readily combines with organic matter (*29*). Thus, organic sulfur enrichment occurs due to reduction of sulfate ion by heterotrophic anaerobic bacterial processes to yield H_2S that subsequently combines with the organic matter. The perhydrous nature of the sulfur-rich vitrinite may be due to preservation of remnant bacterial lipids (*2*) as well as otherwise labile functionalized lipids such as fatty acids, steroids, isoprenoid alcohols or alkenes due to the natural vulcanization of these constituents (*30-31*). The close correlation observed between high organic sulfur and high fluorescence intensity in DH-1494 and DH-7 suggests that the preservation of otherwise labile lipids due to the incorporation of sulfur from H_2S is a reasonable mechanism to explain the origin of certain perhydrous, high sulfur, vitric coals.

Conclusions

The data presented here show that between 0.5 and 1.1% $\overline{R}o_{max}$, fluorometric analysis can be used to identify and correct for the suppression of vitrinite reflectance. What happens to perhydrous vitrinite at higher levels of maturity remains to be investigated. However, if perhydrous vitrinite is likened to liptinite or bitumen, then by the time it reaches the low-volatile stage of coalification it may report a reflectance higher than that of associated vitrinite. For example, AS-2856 (32) shows that sporinite reflectance exceeds vitrinite reflectance above about 1.5% $\overline{R}o_{max}$. Likewise, the reflectance of native bitumen is higher than associated vitrinite below 1% reflectance and higher than associated vitrinite above 1% reflectance (33).

The genetic causes of perhydrous vitrinite that exhibits suppressed vitrinite reflectance and high fluorescence intensity remain to be fully examined; multiple causes are likely. Two different early biogenic alteration mechanisms are suggested here. Besides these heterogeneous anaerobic degradation processes, other mechanisms may also influence the development of suppressed reflectance. Whatever the cause, suppressed reflectance may not be so rare and is significant where present.

Acknowledgments

This paper is a result of research directed towards prediction of the carbonization behavior of New Zealand coals and was funded by the New Zealand Ministry of Commerce. Coal samples were provided by the Coal Corporation of New Zealand, Greymouth Coal Limited, and through a joint University of Canterbury, NZ Ministry of Commerce, and NZ Foundation for Science and Technology, drilling and analytical program.

Literature Cited

1. Thomas, J. R.; Damberger, H. H. *Ill Geol. Surv. Circ. no. 493.* **1976**, 38p.
2. Taylor, G. H.; Lui, S. Y. *Fuel.* **1987**, *66*, 1269-1273
3. Brown, H. R.; Cook, A. C.; Taylor, G. H. *Fuel.* **1964**, *43*, 111-124.
4. Goodarzi, F.; Gentzis, T.; Snowdon, L. R., Bustin, M., Feinstein, S.; Labonte, M. *Mar. Petrol. Geol.* **1993**, *10*, 162-171.
5. Wilkens, R. W. T.; Wilmhurst, J. R.; Russell, N. J.; Hladky, G.; Ellacott, M. V.; Buckingham, C. *Org. Geochem.* **1992**, *18*, 629-640.
6. Lo, H. B. *Org. Geochem,* **1993**, *20,*.653-657.
7. Teichmuller, M. In *Advances in Organic Geochemistry 1973*; Tissot, B., Bienner, F., Eds.; Technip: Paris, 1974, pp 379-408.
8. Teichmuller, M.; Wolf, M. *J. Microscopy.* **1977**, *109*, 49-73.
9. Creaney, S.; Pearson, D. E.; Marconi, L. G. *Fuel,* **1980**, *59*, 438-440.

10. Teichmuller, M. *Fluorescence microscopical changes of liptinites and vitrinites during coalification and their relationship to bitumen generation and coking behavior.* Geol. Surv. of Nordrhein-Westfalen, Krefeld 1982, 119p. (English translation by N. Bostick, Special TSOP pub. no 1., Houston, 1984)

11. Wolf, M.; Wolff-Fischer, E.; Ottenjann, K.; Hagemann, H. W. *Minutes of the 36th ICCP meeting Comm. 3*, Oviedo, Spain, 1983, 14p.

12. Diessel, C. F. K. *Fuel.* **1985**, *64*, 1542-1546.

13. Davis, A.; Rathbone, R. F.; Lin, R.; Quick, J. C. *Org. Geochem.* **1990**, *16*, 897-906.

14. Strauss, P. G.; Russell, N. J.; Bennett, A. J. R.; Atkinson, C. M. In: *Coal Exploration, Proc. of the 1rst Int. Coal Exploration Symp.*; W. L. G. Muir, Ed.; Miller Freeman: San Francisco, 1976, pp.401-443.

15. Quick, J. C. *unpublished Ph.D thesis.* Univ. of Canterbury, 1992, 282p.

16. AS-2486 *Microscopical Determination of the Reflectance of Coal Macerals.* Standards Association of Australia, Standards House, 80 Arthur Street, North Sydney, NSW, 1981, 12p.

17. Lin, R.; Davis, A.; Derbyshire, F. J. *Int. J. Coal Geol.* **1986**, *6*, 215-228.

18. Diessel, C. F. K.; Wolff-Fischer, E. *Int. J. Coal Geol.* **1987**, *9*, 87-108.

19. Quick, J. C.; Davis, A.; Lin, R. *ISS-AIME Ironmaking Conf. Proc.* **1988**, *47*, 331-337.

20. McHugh, E. A. *Adv. Stud. of the Sydney Basin*, 20th Newcastle Symp. Proc., 1986, pp.66-70.

21. Quick, J. C., Davis, A., Glick, D., *Abs. with Prog.* 19th Conf. on Carbon; Univ. Park, PA, 1989, pp.232-233.

22. McHugh, E. A.; Diessel, C. F. K.; Kutzner, R. *Fuel*, **1991**, *70*, 647-653.

23. Veld, H.; Fermont, W. J. J. *Meded. Rijks Geol. Dienst*, **1990**, *45*, 152-170.

24. Taylor, E. McKenzie. *Fuel.* **1926**, *5*, 195-202.

25. Brock, T. D.; Madigan, M. T. *Biology of Microorganisms.* Prentice-Hall: Englewood Cliffs, NJ, 1988, 835p.

26. Belyaev, S. S.; Lein, A. Yu.; Ivanov, M. V., In: *Biogeochemistry of Ancient and Modern Environments*; P. A. Trudinger; M. R. Walter, Eds.; Proc. 4th Int. Symp. Environmental Biogeochem., Canberra, Australia, 1979, pp.235-242

27. Bass Becking, L. G. M.; Kaplan, I. R.; Moore, D. *J. Geol.* **1960**, *68*, 243-284.

28. Robinson, J. M.; *Geology.* **1990**, *15*, 607-610.

29. Casagrande, D. J. In: *Coal and Coal Bearing Strata: Recent Advances*; A. C. Scott Ed.; Geological Society Special Publication no. 32; 1987, pp 87-105.

30. Sinningh Damste, J. S.; Rijpstra, W. I. C.; Kock-Van Dalen, A. C.; de Leew, J. W.; Schenk, P. A. *Geochim. et Cosmochim. Acta.* **1989**a, *53*, 123-141.

31. Sinningh Damste, J. S.; Rijpstra, W. I. C.; A. C.; de Leew, J. W.; Schenk, P. A. *Geochim. et Cosmochim. Acta.* **1989**b, *53*, 143-155.

32. AS-2856 *Coal - Maceral Analysis.* Standards Association of Australia, Standards House, 80 Arthur Street, North Sydney, NSW, 1986, 22p.

33. Jacob, H.; and Hiltmann, W. *Final Report*, Deutsche Gesellschaft Fur Mineralolwissenschaft und Kohlechemie, Project 267; Hamburg, 1985, 54p.

RECEIVED April 15, 1994

Chapter 6

Petrographic and Geochemical Anomalies Detected in Spanish Jurassic Jet

I. Suárez-Ruiz[1], M. J. Iglesias[1], A. Jiménez[1], F. Laggoun-Défarge[2], and J. G. Prado[1]

[1]Instituto Nacional del Carbón (CSIC), La Corredoria, s/n. Apartado 73, 33080 Oviedo, Spain
[2]Unité de Recherche en Pétrologie Organique, Unité de Recherche Associée, 724 du Centre National de la Recherche Scientifique, Université d'Orléans, 45067 Orléans, Cedex 2, France

The Kimmeridgian Spanish jet was studied using petrographic and geochemical methods. Rank parameters obtained from chemical analyses could not be correlated with those determined by petrographic measurements. Explanation of the anomalies could be related to differential absorption in early diagenetic stages, of petroleums generated from the Pliensbachian infraadjacent source-rocks. These oils would only be retained inside ulminite micropores, affecting its reflectance and would give special properties to this coal.

From the point of view of rank studies, vitrinite reflectance (R_0) is one of the parameters most widely used due to its quick and simple obtention and application for determining the organic evolution stages. Moreover it is, in general, recognized as one of the most useful methods for evaluating changes in the organic substances through their geological history.

Although temperature, duration of heating and pressure (the latter in a different way) are the main factors affecting vitrinite reflectance during the coalification process, in the last ten years other factors have been also observed and recently reviewed by Barker (1). These factors may eventually affect the reflectance values and modify the normal growth of reflectance. For this reason, in some cases the suitability of vitrinite reflectance as a rank parameter may be limited and perhaps can not be considered an indicator of organic maturity.

The anomalies in reflectance usually manifest themselves in a reduction of its values when they are compared to those found in profiles or normal evolution series. This is also the case when they are compared to the reflectance values obtained in adjacent sediments with the same diagenetic history. In other cases anomalous values are detected when for the same material vitrinite reflectance disagrees with other rank parameters. The reduction of reflectance just mentioned is known as reflectance suppression (1).

0097–6156/94/0570–0076$08.00/0

In this paper, the relationships between the absorption of oil or petroleum-like substances in the microporosity of huminite (ulminite) and the suppression of its reflectance are studied. Furthermore, the influence of this absorption on coal composition and on other evolution parameters is considered.

The coal chosen, Spanish jet[3], is known due to its special properties, particularly for polishing, and also because it remains unaltered after long exposure to air. This coal has been utilized since XIIth and XIIIth centuries for gem and ornamental purposes. The deposits are found in precise locations in Asturias (Northern Spain), Figure 1, and their geological age corresponds to the Malm (Kimmeridgian) inside detrital facies *(4,5)*. These coals have been formed from drift wood carried away by streams *(6)* and deposited in a transitional marine-continental sedimentary environment.

Previous studies on this coal are scarce and they only refer to its physical characteristics *(5,7)*. On the other hand and taking into account the location area of Spanish jet (the sediments belong to catagenetic stage of evolution), a very low reflectance for this coal was found *(6)*.

Analytical Procedures

The samples were obtained from an underground mine and prepared for petrographic and geochemical analyses.

A petrographic study was carried out on a MPV II Leitz apparatus by means of reflected white light using oil immersion objectives (32x) for the reflectance measurements in accordance with ISO 7404/5 *(8)* procedure. The maceral composition was obtained following the procedures described in ISO 7404/3 *(9)*.

The fluorescence characterization was performed on a MPV III Leitz in water immersion and blue light. Spectral analysis and alteration measurements (after 30 min of exposure) were made by means of U.V. light following the procedure described by Martínez et al. *(10)*. The spectral parameters considered in this work were the quotients QF-535, Q650/500 and the emission flux F.

The organic phase organization, which compliments the petrographic analysis, was carried out by X Ray Diffraction Analysis.

The mineral matter of this coal was analyzed by means of STEM (Scanning Transmission Electron Microscopy) and SEM (Scanning Electron Microscopy) coupled to an EDS (Energy Dispersive Spectrometer). For these analyses, the size of the grains was 2-5 mm. Moreover, the crystallized phases associated with this vitrain were determined by X-Ray Diffraction in a powdered sample.

The chemical characterization (Proximate and Ultimate Analyses, Sulfur Forms and Calorific Value) was made in accordance with the international standard procedures, ISO-589, ISO-1171, ISO-562, ISO-925, ISO-157, ISO-1928 *(11-16)*. The C, H, N and S_{total} content was determined using a LECO CHN 600 and LECO SC 132

---

[3]"Jet is formed from drifted wood which has been secondarily impregnated with bitumen from the surrounding environment leading to an abnormally low reflectance, strong fluorescence and uniform physical properties" *(2)*.
"Jet is drift wood (stems and branches), embedded in Jurassic oil shales, which is impregnated with secondarily bitumen. They outcrop in England, France and Southern Germany" *(3)*.

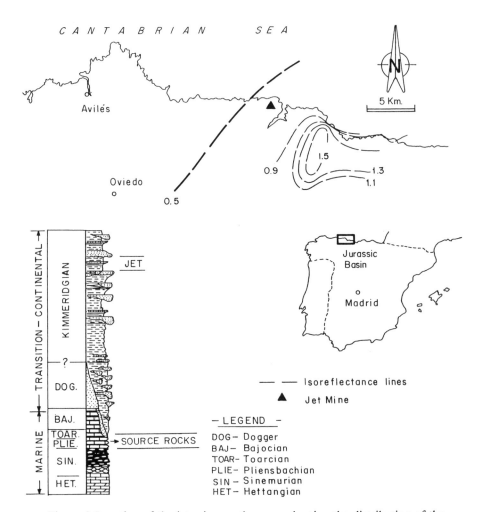

Figure 1.Location of the jet mine on the map, showing the distribution of the rank of the organic matter from the Jurassic according to the vitrinite reflectance data (6). Strat-column of Jurassic sediments from Asturias (Northern Spain) (38).

apparatus, and the oxygen was calculated by difference.

The oil and tar yields of the raw material were obtained through pyrolysis techniques: Gray-King Assay -ISO 502 (*17*)-, Rock-Eval -performed in accordance with Espitalié (*18,19*) and thermogravimetry in the conditions described in a previous paper (*20*). Furthermore, parameters affecting type, composition and evolution of this coal, obtained from Rock-Eval pyrolysis, were also considered.

The functional groups of Spanish jet and its oil -obtained through pyrolysis- were studied by Fourier Transform Infrared Spectroscopy (FTIR). The solid samples were prepared using the traditional procedure (*21,22*), the ratio being coal:KBr of 1:100 and more than one pellet was prepared. The oil was dissolved in dichloromethane and dropped gently on a KBr pellet and the solvent evaporated. All absorbance spectra were recorded on a Perkin Elmer 1750 spectrometer co-adding 25 interferograms obtained at a resolution of 4 cm^{-1}. Each spectrum was corrected for scattering and then calibrated to 1 mg (daf)/cm^2.

The soluble organic fraction was obtained according with the extraction procedure described by Blanco *et al.* (*23*). Thus, a weighed amount of ground coal (< 0.2 mm) and a measured amount of chloroform (1 g of solute/20 mL of solvent) were placed in a glass flask. The closed glass was put into an ultrasonic bath kept at 25 °C for 2 h. The solution was then filtered and the residue was washed until the filtrate became totally clear. The extract was isolated by means of solvent evaporation at atmospheric pressure. It was then fractionated by liquid chromatography and the content of saturated and aromatic hydrocarbons as well as resins and asphaltenes was determined in order to obtain the mass balance.

The textural analysis of this coal was carried out through the measurement of real and apparent densities as well as porosity. For determining apparent density mercury porosimetry was used. The real density was determined using a glass pycnometer. The calibration gas used was helium (pycnometric fluid). Porosity was calculated through the density data.

Results.   From the macroscopic point of view, Spanish jet has homogeneous appearance without structure nor texture of vegetable origin. It is hard, compact, black, bright and glassy. Its cubic and/or conchoidal fracture has straight and sharp edges. It has no cracks, it is stainless and also light-weight and easy to polish. Furthermore, minerals are not observed and, in comparison with other coals and vitrains of low rank, it is unaltered after exposure to air. The colour of the mark made on porcelain is brown. The Vickers' microhardness (*5*) is 85 Kp/cm^2, which is equivalent to a hardness of 2.35 Mohs' scale.

Optically, the maceral group of huminite/vitrinite (98.8% vol.) is the only organic component of this coal and its mineral matter content is very low (1.2% vol.). Inside the huminite/vitrinite, ulminite is the highest component (85.5% vol. of the vitrinite total content) followed by phlobaphinite (14.5% vol.) as shown in Table I.

Ulminite, darker than phlobaphinite, has a random reflectance of 0.39% and it looks like a vegetable tissue of well-defined cellular walls and cavities. On the other hand, phlobaphinite appears as the filling of cellular cavities and segregations mixed with the ulminite. Phlobaphinite shows different topographic drawings depending on the section of the coal observed. Its reflectance (0.72%) is in all cases higher than that found in the ulminite (see Table I).

Table I. Maceral Analysis and Reflectance Measurements of Spanish Jet.

	Maceral Analysis (%vol.)	Reflectance Analysis (%R_0)
Organic fraction	98.8	
Mineral fraction	1.2	
Ulminite (m.m.f.)	85.5	0.39 (0.04)[a]
Phlobaphinite (m.m.f.)	14.5	0.72 (0.06)[a]

[a]Values in parentheses are standard deviation for the reflectance determinations.

It must also be pointed out that other components like resinite or traces of its impregnation have never been observed in this coal.
The results of fluorescence microspectrometry are given in Table II. Only one of the maceral components mentioned, ulminite, fluoresces. Its spectrum has a very low intensity (F: 2.9 a.u.) and it is shifted to high wavelengths (λ_{max}.: 598 nm). The high values found for the parameters QF-535 and Q650/500 (3.05 and 1.38, respectively) are in good agreement with the above mentioned spectral data. The fluorescence alteration of the ulminite was also studied and the data are given in the same Table. It is positive during the first five minutes (slight fading +) and then constant to the end of the analysis. The chromatic derivative QF shows the same trend.

Table II. Fluorescence Properties of Spanish Jet.

	Spectral Analysis			Spectral Alteration	
	F (a.u.)	λ_{max}. (nm)	QF-535	F	QF
Ulminite	2.09	598	3.05	positive (5') constant(>5')	constant
Phlobaphinite	Non fluorescent				

The mineral matter detected by means of SEM/EDS is scarce. Gypsum and/or anhydrite (with some substitution of Ca by K), clays (mainly silicates of Fe and Al), traces of carbonates and some pyrite were the mineral phases identified.
In relation to the organization of aromatic and/or aliphatic compounds, the XRD analysis shows, in general, an organic phase with a small degree of structural order. The average value of aromatic layer spacing (d_{002} = 3.82 Å) is very close to that found in lignites.

Table III shows the results of the chemical analysis. The very high carbon content for this coal, close to that corresponding to bituminous coals (2), must be taken into account. Spanish jet also shows a high calorific value and hydrogen content. These values are not in agreement with the volatile matter and ulminite reflectance. Furthermore, its moisture content is too small compared to other coals inside the rank from lignite to high volatile bituminous coal (2) and, in general, compared to the rank scale of coals. The low ash content found in this coal agrees with the low amount of mineral matter determined by means of optical procedures. These results suggest that this coal has a very pure organic composition, and few minerals.

Table III. Chemical Characterization of Spanish Jet.

Proximate Analysis (%)		Sulfur Forms[b] (%)	
Moisture	2.91	S_{total}	1.03
Ash[a]	1.08	$S_{org.}$	0.95
Volatile Matter[b]	54.92	$S_{pyr.}$	0.05
Ultimate Analysis[b] (%)		$S_{sulph.}$	0.03
C	84.81		
H	5.87	Calorific Value[b,c]	35.21
$O_{diff.}$	7.47		
N	0.90		
Atomic Ratio[b]			
O/C	0.07		
H/C	0.83		

[a]Dry basis
[b]Dry ash free
[c]MJ/Kg

In Figures 2 and 3 the correlations between different evolution parameters of coals for increasing rank are given. These figures clearly illustrate all the above mentioned discrepancies found in Spanish jet. Thus, the shift found for this coal in Figure 2a points to its too high carbon content compared to the ulminite reflectance value gives in Table I. It can be also said that the %C is high when reflectance on the phlobaphinite is considered. However, this %C is more or less in agreement with oxygen content (see Figure 3). The volatile matter content is high showing a normal

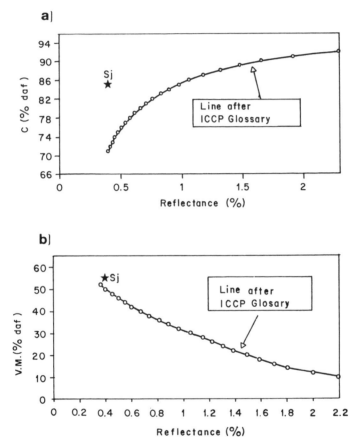

Figure 2.Spanish jet location in a) the vitrinite reflectance/C and b) vitrinite reflectance/volatile matter diagrams established for a coal set of increasing rank (*39*).

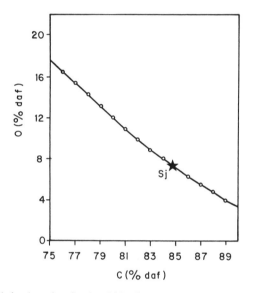

Figure 3. Spanish jet location in the C/O diagram established for a coal set of increasing rank (*40*).

correlation with ulminite reflectance but not with phlobaphinite reflectance (Figure 2b).
The high carbon and hydrogen contents and consequently the high atomic ratio H/C shift the position of Spanish jet in van Krevelen' diagram (Figure 4a) with respect to the tendency of humic coals (kerogen type III). Nevertheless, this position is close to certain special coals (24,25), which have been described as perhydrogenated coals with anomalous properties.
The results obtained from the application of the different pyrolysis techniques used in this work are given in Table IV.

Table IV. Extraction and Pyrolysis Data of Spanish Jet.

Soluble Fraction(%)[a]		Rock-Eval		Gray-King	
Extraction yield	0.21	S_1[b]	1.95	Oil Yield[f]	316.72
Saturated HC	8.76	S_2[b]	354.71		
Aromatic HC	13.87	$T_{max.}$[c]	422		
Resins, Asphaltenes	77.37	O.P.T.	0.01		
		TOC[d]	95.13		
		HI[e]	372		

[a]Chloroform was used as solvent
[b]mg HC/g rock
[c]°C
[d]%weight
[e]mg HC/g TOC
[f]mg/g rock

With respect to Rock-Eval pyrolysis, the low value of $T_{max.}$, considered as an organic maturity parameter, points to an immature organic substance. This value could be related to ulminite reflectance. The high TOC and HI values were expected taking into account the elemental analysis (see %C and %H in Table III). However, HI, $T_{max.}$ values and hence the anomalous position of the Spanish jet in the $T_{max.}$/HI diagram (see Figure 4b) compared with a set of coals of increasing rank (26) point also to an immature organic matter close to a type II kerogen rather than a kerogen of type III, which is prone to generate hydrocarbons. This high petroleum potential is confirmed by means of the value of S_2 (Rock-Eval) and oil yield (Gray-King Assay) given in Table IV. Both parameters have similar values and they are too high when they are compared with that found in other coals (27,28),and especially in coals made up only of vitrinite. Moreover, the oil yield obtained from Spanish jet is higher than that found in some immature kerogens of types I and II (29). However, the oil yield agrees with HI, H and TOC contents as well as with weight loss (about 45% of the

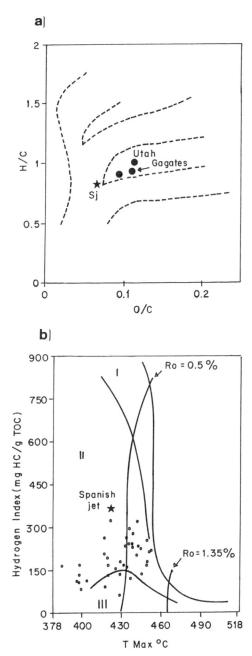

Figure 4. Location of the Spanish jet in a) van Krevelen'and b) $T_{max.}$/HI diagrams in relation to other coals (*24-26, 29*).

total weight loss) obtained at the end of the pyrolysis carried out by means of thermogramivetry (20). Furthermore, the thermal behaviour of Spanish jet is very close to type I kerogens (30).

Table IV also shows the extraction yield in chloroform of this coal as well as the composition of the extract. As can be observed a very small amount of extract (0.21%) was obtained. This extract is made up of heavy and polar compounds (resins and asphaltenes, 77.37%) whereas the concentration of saturated hydrocarbons is small (8.76%). This low saturated hydrocarbon content agrees with the relatively low S_1 value (see Rock-Eval data in Table IV) which is typical of immature materials containing heavy compounds (31).

Another important aspect in coal characterization is the study of the functional groups. This study was preformed by means of FTIR and Figure 5 shows the spectra of Spanish jet and its oil as well as the Utah jet spectrum taken from the literature (24). In the aromatic C-H stretching region of the raw material spectrum a significant absorption at ca. 3030 cm^{-1} can be observed. Taking into account that this mode shows a consistent tendency towards increasing values with increasing rank (31,32) the intensity of the band at 3030 cm^{-1} observed in Spanish jet might be more consistent with its carbon content than ulminite reflectance. Furthermore the insignificant absorption in the region of C=O vibration modes indicates a bituminous rank of this coal rather than lignite (33), i.e. the evolution stage of Spanish jet might be higher than that corresponding to the transition diagenesis/catagenesis.

On the other hand as in most coals, three principal bands between 900 and 700 cm^{-1} (out-of-plane vibration of aromatic C-H bonds) centered at about 870, 815 and 750 cm^{-1} can be observed. It must borne in mind that these bands undergo shifts in frequency as function of rank, particulary the mode characteristic of lone C-H groups near 870 cm^{-1} shifts to lower frequency in coals with a lower carbon content (32,35). Furthermore, this region of the spectra can be used to follow the changes in aromatic substitution (22) with increasing rank. In general, the degree of substitution is high in coals with a low carbon content and it is reduced as the carbon content increases. Nevertheless, in most coals, the band at 870-860 (750) cm^{-1} is the highest (smallest) of the three bands found between 900-700 cm^{-1}. In spite of this, the intensity of the band at 870 cm^{-1} in Spanish jet spectrum is slightly lower than the band centered at about 750 cm^{-1} whereas the most intense is the band at 815 cm^{-1}. Moreover, a clear absorption at ca. 1490 cm^{-1} can also be observed. This band, which is attributed to certain kinds of substitution or condensation of aromatic systems (36), is usually detected in coals only as a shoulder (22) From these results, aromatic systems with a unusual substitution could be expected. In relation to these observations, the similarity between the spectra of Spanish and Utah jets must be underlined (see Figure 5).

Figure 5 also shows the spectrum of the oil obtained by means of pyrolysis (Gray-King Assay). A high similarity between this spectrum and the one corresponding to the raw material can be observed, especially in the above mentioned regions. From this observation, these materials might be composed of similar structures.

Finally, the results of textural characterization are given in Table V. The apparent density corresponds to the one described for sub-bituminous and bituminous coals whereas the real density is slightly lower than that for these coal ranks. However, its porosity is unusually small in comparison with the other parameters found for this

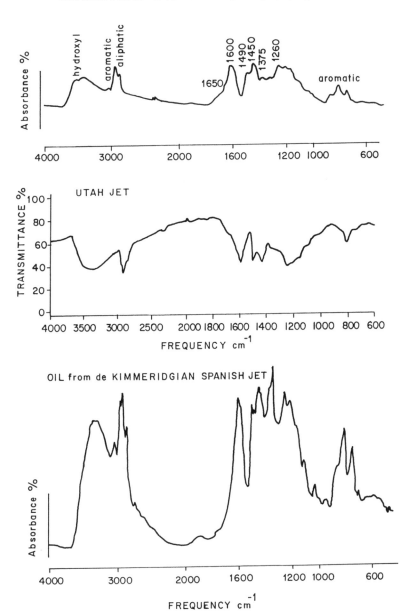

Figure 5.FTIR spectra of Spanish jet, its oil and Utah jet (*24*).
Reproduced with permission from reference 24. Copyright 1968 American
Association for the Advancement of Science.

coal and also with the porosity values of other coals. The absorption of certain substances in the porosity of this coal is the most probable explanation of this feature. Furthermore, the existence of these substances allows the special characteristics found in Spanish jet to be explained.

Table V. Textural Properties of Spanish Jet in Relation to Other Vitrains.

Sample	r.d.[a]	a.d.[a]	P.
Cretaceous (Albien)[b]	1.27	1.12	106
Spanish Jurassic jet	1.23	1.17	42
French Carboniferous Stephanian[b]	1.37	1.17	125
Spanish Carboniferous Stephanian[b]	1.38	1.18	123
French Carboniferous Stephanian[b]	1.34	1.16	116
Spanish Carboniferous Stephanian[b]	1.34	1.20	87

r.d.: real density in g/cc   [a]Dry Basis
a.d.: apparent density in g/cc   [b]Unpublished results
P.: Porosity in mm^3/g

Discussion From the above results, no well defined correlations between composition, properties and evolution parameters can be established. Likewise, they are not expected, taking into account the characteristics described for "normal" coals. The ulminite reflectance value (Table I), fluorescence parameters (Table II), volatile matter and hydrogen content (Table III), as well as Rock-Eval data ($T_{max.}$, HI, S_1 in Table IV) and the low level of organic phase organization locate Spanish jet in an immature diagenetic stage and its high hydrogenation (see Tables III and IV and Figure 4) agrees with its oil potential (see Table IV). However, its high carbon content and TOC (Tables III and IV, respectively), the phlobaphinite reflectance and FTIR results can not be explained. These latter data show the existence of compounds with structures close to those found in coals of higher rank like high volatile bituminous coals.

The hypothesis proposed in this paper to explain the anomalies and discrepancies found in Spanish jet is based on its impregnation by oil or petroleum-like substances in the early diagenetic stages. Taking into account the vitrinitic composition of this coal, these substances must be generated and migrated from other sediments.

The results found in the study of the porosity confirm that the microporous system of this coal is probably filled with an organic material. Nevertheless, the most important evidence in the above explanation is the location of the Spanish jet. This coal is found on top of the Pliensbachian sediments, which are considered as a source-rock in this area of the Northern Spain (*6*). From the point of view of organic maturity these Pliesbachian sediments are in the oil window or oil generation stage (*37*). In almost all the sedimentary series over the Pliensbachian, petroleum impregnations have been found. Thus, the supra- and infra-adjacent sedimentary levels of this coal show moldic porosity, which is full of petroleum-like substances.

However, impregnation of the Spanish jet has not been uniform. In fact, a differential absorption depending on the vitrinite component (ulminite or phlobaphinite) is proposed. Thus, the petroleum-like substances would only be retained inside ulminite micropores but not inside phlobaphinite probably because it has a smaller porosity than ulminite and also because it is more resistent to impregnation by oils.

Ulminite impregnation could have caused the suppression of its reflectance during the subsequent coalification. However, the reflectance of the unimpregnated component, phlobaphinite, could increase during this process until the high volatile bituminous coal rank is reached. Therefore, phlobaphinite reflectance would be in agreement with the rank described for the location area of Spanish jet (Figure 1) and it could be the probable degree of evolution reached for this coal.

However, the characteristics and properties of this coal would be determined principally by the impregnated ulminite because it is the main component of this coal (84.5% of the total amount of organic and mineral matter). For this reason, impregnation would be responsible for the high values of %C, TOC, MV, %H, HI, calorific value and petroleum potential found in the Spanish jet. The impregnations could also produce the fluorescence properties of ulminite. Likewise, some of the different characteristics found in ulminite fluorescence (slight fading +) in comparison with the secondary fluorescence in "normal" vitrinites could be due to the different source of impregnation in this coal. It must be borne in mind that the secondary fluorescence of vitrinites is induced by bitumen generated by the coal itself (*26*) whereas in Spanish jet the bitumen/oil has come from other sediments.

Finally, the S_1 value of Rock-Eval, the aromatic modes of vibration observed in FTIR and the composition of the extract could indicate that the hydrocarbonated fraction in the microstructure of ulminite corresponds to the heavy fraction (polar components). These compounds must be the responsible for the supression of reflectance as well as of the anomalies detected in the Spanish jet.

Conclusions. From the results obtained in this work, the following conclusions can be inferred:

* Although the characteristics of the coal studied allow it to be included in the coal group denoted as "jet", it is evident that the genesis and process giving rise to these special properties are different in Spanish jet to those in other jets. Thus, whereas the presence of resinitic compounds confers the high degree of hydrogenation to some of these coals, the features found in Spanish jet are clearly due to its impregnation by oil migrated from other sediments.

* This impregnation takes place probably in the initial diagenetic stages and it has not been uniform. The vitrinitic/huminitic components of this coal have undergone differential impregnation. Thus, the hydrocarbons have been retained only inside ulminite micropores, and this comfirms the genetic differences between huminite/vitrinite macerals.

* The hydrocarbon impregnation/absorption produces reflectance suppression in ulminite, and the initial differences in reflectance between the two components are increased in the subsequent coalification. For this reason, it is possible to identify phlobaphinite after the diagenetic/catagenetic transition when usually it is only possible in low ranks.

* Another proof of this impregnation is the fluorescence of ulminite. This confirms that vitrinite fluorescence and reflectance suppression are caused not only by the bitumens generated from the coal but also by oils migrated from other sediments.

* The retention of the heavy fraction (polar components) of hydrocarbons in the microporosity of ulminite is also responsible for the anomalies detected in the composition of this coal and the discrepancies between rank parameters.

* Phlobaphinite reflectance might be the closest value for determining the rank of this coal because it is similar to the degree of evolution found in the location area of Spanish jet.

* It is evident that the impregnated coals are important in two ways. First, they are an indication of the areas which are generating or have generated hydrocarbons and consequently they might also indicate the existence of petroleum in adjacent sediments. Furthermore, these coals became fuels with a high calorific value and an important petroleum potential.

* The identification of vitrinites with unusually low reflectance and the causes of this suppression are of widespread interest for determining the suitability of vitrinite reflectance as a rank parameter and also because it represents processes which have taken place in a specific sedimentary basin.

Acknowledgments. This work was sponsored by EC, Contract n° 7220-EC-757.

Literature Cited
(1)   Barker, Ch.E. *The Soc. for Org. Petrology, Newsletter* **1991**, *8*, 8.
(2)   Stach, E.; Mackowsky, M. Th.; Teichmüller, M.; Taylor, B.H.; Chandra, D.; Teichmüller, R. *Coal Petrology*; 3th Edition; Gebruder Borntraeger: Stuttgart, Berlin, 1982; 535 pp.
(3)   Teichmüller, M. *Intern. Journal of Geology* **1992**, *20*, 1.
(4)   Suárez-Vega, L.C. *Estratigrafía del Jurásico de Asturias*; Publicaciones del CSIC: Madrid, Spain, 1974; Vol. 1 and 2.
(5)   Campon, E.; Fernández, C.J.; Solans Huguet, J. *Trabajos de Geología, Univ. de Oviedo* **1978**, *10*, 161.
(6)   Suárez-Ruiz, I. *Caracterización, clasificación y estudio de la materia orgánica dispersa (MOD) en el Jurásico de Asturias y Cantabria*; Ph.D.: University of Oviedo, Spain, 1988; Vol. 1 and 2.
(7)   Solans Huguet, J.; Campon, E.; Fernández, C.J. *A.E. de Geología* **1980**, *12*, 5.
(8)   *Methods for the petrographic analysis of bituminous coal and anthracite.- Part 5: Method of determining microscopically the reflectance of vitrinite.*; International Standard ISO 7404/5, 1th Edition, 1984; 13 pp.
(9)   *Methods for the petrographic analysis of bituminous coal and anthracite.- Part 3: Method of determining maceral group composition*; International Standard ISO 7404/3, 1th Edition, 1984; 4 pp.

(10) Martínez, L.; Pradier, B.; Bertrand, P. *C.R. Acad. Soc. Paris* **1987**, *304 Série III/9*, 441.

(11) *Hard coal. Determination of total moisture*; International Standard ISO 589, 2th Edition, 1981; 6 pp.

(12) *Solid mineral fuels. Determination of ash*; International Standard ISO 1171, 2th Edition, 1981; 2 pp.

(13) *Hard coal and coke. Determination of volatile matter content*; International Standard ISO 562, 2th Edition, 1981; 5 pp.

(14) *Solid mineral fuels. Determination of carbon dioxide content. Gravimetric method*; International Standard ISO 925, 2th Edition, 1980; 3 pp.

(15) *Hard coal. Determination of forms of sulphur*; International Standard ISO 157, 1th Edition, 1975; 10 pp.

(16) *Solid mineral fuels. Determination of gross calorific value by the calorimetric bomb method and calculation of net calorific value*; International Standard ISO 1928, 1th Edition, 1976; 14 pp.

(17) *Coal. Determination of coking power. Gray-King Assay.*; International Standard ISO 502, 2th Edition, 1981; 9 pp.

(18) Espitalié, J.; Madec, M.; Tissot, R.; Mening, J.J.; Leplat, P. *Proc. 9th Ann. Offshore Techn. Conf.*, Houston, Tx., **1977**, 439.

(19) Espitalié, J.; Laporte, J.L.; Madec, M.; Marquis, F.; Leplat, P.; Paulet, J.; Boutefeu, A. *Rev. de l'Instit. Français du Pétrole* **1977**, *32*, 23.

(20) Jiménez, A.; Fuente, E.; Laggoun-Defarge, F.; Suárez-Ruiz, I. *IX Colloque Internat. des Petrographes Organiciens Francophones*; June, 1993: Pau, France; 3 pp.

(21) Solomon, P.R.; Hamblen, D.G.; Carangelo, R.M. In *Coal and Coal Products: Analytical Characterization Techniques*; Fuller Jr., E.L., Ed.; ACS Symposium Series No. 205; American Chemical Society: Washintong, DC, 1982; 77-131.

(22) Painter, P.; Starsinic, M.; Coleman, M. In *Fourier Transform Infrared Spectroscopy*; Ferraro, J.R.; Basile, L., Eds.; Academic Press: Orlando, 1985; Vol. 4, 169-241.

(23) Blanco, C.G.; Prado, J.G.; Guillén, M.D.; Borrego, A.G. *Org. Geochem.* **1992**, *18*, 313.

(24) Traverse, A.; Kolvoord, R.W. *Science* **1968**, *159*, 302.

(25) Petrova, R.; Mincev, D.; Mikolov, ZDR. *Internat. Journal of Coal Geology* **1985**, *5*, 275.

(26) Teichmüller, M.; Durand, B. *Internat. Journal of Coal Geology* **1983**, *2*, 197.

(27) Teerman, S.C.; Hwang, R.J. *Org. Geochem.* **1991**, *17*, 749.

(28) Verheyen, T.V.; Johns, R.B.; Espitalié, J. *Geochim. Cosmochim. Acta* **1984**, *48*, 63.

(29) Espitalié, J.; Deroo, G.; Marquis, F. *Revue de l'Istit. Français du Petrole* **1985**, *40*, 755.

(30) Durand-Souron, C. In *Kerogen: Insoluble Organic Matter from Sedimentary Rocks*; Durand, B., Ed.; Editions Technip: Paris, 1980; 519 pp.

(31) Snowdon, L.R. *Bull. Canad. Petr. Geol.* **1984**, *32*, 327.

(32) Brown, J.K. *J. Chem. Soc., London* **1955**, 744.

(33) Kuehn, D.W.; Snyder, R.W.; Davis, A.; Painter, P.C. *Fuel* **1982**, *61*, 682.

(34) Cooke, N.E.; Fuller, O.M.; Gaikwad, R.P. *Fuel* **1986**, *65*, 1254.

(35) Riesser, B.; Starsinic, M.; Squires, E.; Davis, A.; Painter, P.C. *Fuel* **1984**, *63*, 1253.

(36) Elofson, R.M. *Can. J. Chem.* **1957** *35*, 926.

(37) Suárez-Ruiz, I.; Prado, J.G. *NATO Advances Study Institute: Composition, Geochemistry and Conversion of Oil Shales. Short Course*; July, 1993: Turkey; Abstract Booklet 86 pp.

(38) Valenzuela Fernández, M. *Estratigrafía, sedimentología y paleogeografía del Jurásico de Asturias*; Ph.D.: University of Oviedo, Spain, 1988; Vol. 1 and 2.

(39) *International Handbook for Coal Petrology*; ICCP Glossary, 2th Ed., CNRS: Paris, 1975, Vol. 1,2 and 3.

(40) Hickling, H.G.A. *Proc. S.W. Inst. Eng.* **1931**, *46*, 911.

RECEIVED June 9, 1994

Chapter 7

Reflectance Suppression in Some Cretaceous Coals from Alberta, Canada

Thomas Gentzis[1] and Fariborz Goodarzi[2]

[1]Alberta Research Council, Coal and Hydrocarbon Processing,
One Oil Patch Drive, Devon, Alberta T0C 1E0, Canada
[2]Geological Survey of Canada, Institute of Sedimentary and Petroleum
Geology, 3303 Thirty-Third Street Northwest, Calgary,
Alberta T2L 2A7, Canada

The unusual petrological characteristics of the Mannville coals from Alberta have been investigated using optical microscopy, Rock-Eval pyrolysis and biomarker geochemistry. A strong influence of the brackish-water depositional environment, as revealed by the extremely high boron and total sulphur contents, on the fluorescence of vitrinite in these coals is suspected. This may have resulted in a significant suppression of the measured reflectance. Some of the coals are highly-enriched in hydrogen, even at high levels of thermal maturity. There is a good correlation between Tmax and %Ro, max but the correlation between HI and %Ro, max is moderate. Contribution from liptinite macerals to vitrinite fluorescence is minimal and it is suggested that the coals have absorbed bitumen, which is partly autochthonous (formed *in situ*) and partly allochthonous (not formed *in situ*). The study also points out some of the problems associated with the use of vitrinite reflectance in regional thermal maturity studies and suggests a multidisciplinary approach in solving these problems.

Use of Vitrinite Reflectance and Introduction to Study Area

Vitrinite reflectance (Ro) is considered to be one of the most reliable and powerful methods used by coal and organic petrologists in assessing the thermal maturity of coals and rocks in sedimentary basins (1). Vitrinite reflectance covers the entire thermal maturity range, from early diagenesis to metamorphism, and records only the maximum temperature to which organic matter has been subjected. Although vitrinite reflectance measured either on vitrinite group macerals in coal or on the equivalent Type-III kerogen dispersed in sediments can solve numerous geological problems, some limitations do exist concerning the use of this important parameter.

0097–6156/94/0570–0093$08.00/0
Published 1994 American Chemical Society

These are related to: (a) variations in vitrinite reflectance according to matrix lithology and depositional environment (organic facies), and (b) suppression of vitrinite reflectance in coal or sediments, usually associated with a high content of exinite and liptinite macerals (such as an oil shale). Jones and Edison (2) and Price and Barker (3) recognized that Ro is suppressed in sedimentary rocks enriched in structured or amorphous liptinitic and exinitic organic matter. Hutton and Cook (4) observed that reflectance of vitrinite in an oil shale in Australia is considerably suppressed and that band vitrinite (telocollinite) has higher reflectance than matrix vitrinite (desmocollinite).   The work of Kalkreuth (5) on coals from British Columbia, Canada, indicated that there is a direct relationship between vitrinite reflectance and volume concentration of liptinite and exinite macerals. Snowdon et al. (6) detected a suppression of vitrinite reflectance in liptinite-rich coals, termed 'needle coals', of Mesozoic-age in British Columbia. A study of Carboniferous-age oil shales from Arctic Canada by Goodarzi et al. (7) showed that vitrinite reflectance can be suppressed by as much as 0.4%.   In addition, preliminary studies on Cretaceous Mannville Group coals from the Alberta Plains indicate that the Ro suppression is real and in the order of 0.2 to 0.4% (Pearson, D.E., personal communication, 1988.) Variations in vitrinite reflectance with matrix lithology have been reported by Mukhopadhyay (8) and Goodarzi et al. (9). Possible mechanisms proposed include a differential thermo-catalytic effect and thermal conductivity of the host matrices, absorption of migrated bituminous substances by the vitrinite matrix during catagenesis, or differences in vitrinite chemistry, which is a reflection of its depositional environment. Reflectance of vitrinite is a function of its hydrogen content. Vitrinites of sapropelic coals, deposited in a lower delta plain to marginal marine environment show an increased fluorescence intensity (perhydrous), a higher hydrogen and nitrogen content and a suppressed reflectance (10-11).

In order to discuss the factors affecting Ro suppression in coals, we have chosen a series of petrologically unusual coals from the western Canada sedimentary basin, in the central Alberta Plains (Figure 1).   Hacquebard (12) measured the reflectance of Cretaceous Mannville Group coals in Alberta (%Ro= 0.41 to 1.58) and demonstrated a rank increase from east to west, toward the axis of the Alberta Syncline. Osadetz et al. (13) created a large database on the reflectance of Mannville Group coals in the Alberta Plains and Foothills as well as of Mannville-equivalent coals in northwestern Alberta (Gething Formation).

The Lower Cretaceous strata in the central and southern Plains and Foothills of Alberta are part of a clastic wedge deposited in a foreland basin, which received synorogenic clastic detritus during the emergence of the Cordillera (14). Banerjee and Goodarzi (14) determined the boron (B) and sulphur (S) contents of a suite of Mannville Group coals in southern Alberta. They found B to range from 18 to 1144 ppm, with the majority having greater than 100 ppm, and S having a range between 0.56 and 23.5%. Based on the work of Goodarzi and van der Flier-Keller (15), it is evident that the Mannville coals were deposited in an environment strongly influenced by brackish waters, in agreement with other sources of evidence. As a result, it is anticipated that the vitrinite present in these coals would not be similar to 'humic', oxygen-rich and non-fluorescing vitrinite, rather it will be of the 'sapropelic', hydrogen-rich, oxygen-poor and fluorescing type.

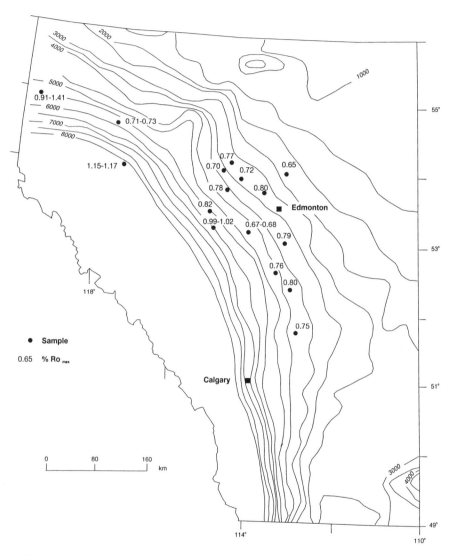

Figure 1. Map of Alberta showing the vitrinite Ro, max values of Mannville coals. Also shown are the isopach contour lines to the top of the Mannville Group.

Experimental

Twenty six Mannville Group coal samples were taken from sixteen drillholes (Table I). The samples were crushed to 850 microns, mounted in resin epoxy, ground and polished according to standard techniques. Maximum reflectance in oil (n oil = 1.518 at 546 nm) of vitrinite was measured using a Zeiss MPM II microscope, an Epiplan-'Neofluor' x40 (N.A. 0.90) objective. Macerals were identified, following the maceral classification recommended by the International Committee for Coal Petrology (16). Fifty reflectance measurements (Ro, max) were taken on each sample and the maceral composition was determined under ultraviolet light only on 500 points, using an automatic point counter. This was done to estimate the amount of fluorescing vitrinite and total exinite and liptinite macerals in the coals. All samples were subjected to Rock-Eval/TOC pyrolysis (Table I), according to the method described by Espitalie et al. (17).

A number of coals were analyzed for a separate study using off-line flash pyrolysis-gas chromatography on a CDS Model 120 Pyroprobe apparatus. The pyrolysis products were extracted at room temperature with n-pentane and injected into a capillary gas chromatograph. Selected coals were extracted with azeotropic chloroform:methanol (87:13) and fractionated into saturates, aromatics, asphaltenes and NSO's using a silica gel/alumina column. The saturate fraction was analyzed using a hybrid VG 70SQ gas chromatograph-mass spectrometer, equipped with multiple ion detection for nominal masses of selected compounds such as steranes and terpanes.

Organic Petrology

Two samples from the Simonette well at 2196-2200 m and 2215-2218 m (Table I) are dominated by vitrinite (85 and 86 vol.%, respectively), mostly fluorescing telo- and desmocollinite (Figure 2a) having Ro, max of 0.71 and 0.73%. Inertinite content is 10% in both samples and consists of semifusinite, fusinite, inertodetrinite (Figure 2a), macrinite and granular micrinite. Liptinite is 5 and 4% respectively, comprising sporinite (Figure 2a), cutinite, resinite, liptodetrinite, exsudatinite (Figure 2b) and fluorescing corpocollinite (Table II). Dinoflagellate cysts and bitumen staining are evident in the shaley fragments of the two samples as is framboidal pyrite. Six coal samples were taken from core from the Pembina well, over a narrow interval (1651-1653.7 m) (Table I). Maximum vitrinite Ro ranges from 0.67 to 0.70%, total vitrinite content is between 55 and 98 vol.%, with the exception of the sample at 1651 m (25%) (Table II). Most vitrinite is fluorescing (Figure 2c) and cell structure is occasionally visible. Liptinite ranges from 2 to 13 vol.%, consisting of sporinite, cutinite, liptodetrinite (Figure 2c), resinite and granular exsudatinite (Figure 2d). Inertinite is low in some samples but increases to 70 vol.% in the sample at 1651 m. Granular micrinite is also present in the sample.

Mannville coal in three of the four Cherhill wells have Ro, max ranging from 0.75% at 1394 m, to 0.91% at 1418 m. Coal in the fourth well has the highest Ro, max measured in this study (1.47%) at a depth of 2579 m (Table I). The maceral composition is shown in Table II. Coal from the Stony Plain well has Ro, max of 0.80% at 1154 m (Table I) but most of the vitrinite fragments are oxidized and

Table I. Sample Depth (m), % Ro, max and Rock-Eval Pyrolysis Data (in duplicate) of the Mannville Group Coals

Sample No.	Depth (m)	%Ro, max	Tmax	S1	S2	S3	PI	Sample No.	Depth (m)	%Ro, max	Tmax	S1	S2	S3	PI
1	2196-2200	0.71	438	2.54	141.17	5.88	0.02	14	1345.0	0.78	458	0.00	2.89	2.46	0.00
1			438	2.59	120.92	4.81	0.02	14			456	0.15	2.76	4.15	0.05
2	2216-2218	0.73	437	1.96	85.88	5.68	0.02	15	1245.0	0.72	445	0.63	9.62	2.40	0.06
2			436	2.22	90.74	5.74	0.02	15			443	0.50	8.75	2.50	0.05
3	1651.0	0.67	441	15.51	144.28	7.34	0.10	16	1806.0	0.82	447	0.90	17.42	2.42	0.05
3			441	29.38	146.73	4.48	0.17	16			445	0.88	25.73	3.97	0.03
4	1651.8	0.67	439	6.06	177.21	3.27	0.03	17	865.0	0.65	442	0.18	9.81	3.70	0.02
4			440	6.19	177.14	2.22	0.03	17			443	0.18	9.62	5.00	0.02
5	1652.2	0.67	438	7.64	191.56	5.88	0.04	18	1340.0	0.79	443	3.33	63.50	4.03	0.05
5			437	6.29	151.48	4.81	0.04	18			444	2.98	46.49	2.45	0.06
6	1652.5	0.69	440	7.75	190.81	6.12	0.04	19	1310.0	0.70	440	0.00	5.93	1.27	0.00
6			442	7.40	182.40	5.60	0.04	19			448	0.00	5.86	1.72	0.00
7	1653.1	0.70	433	7.96	204.25	5.55	0.04	20	1209.0	0.77	443	1.60	43.00	9.80	0.04
7			433	9.50	211.50	2.83	0.04	20			446	1.20	41.40	8.60	0.03
8	1653.7	0.68	438	11.81	174.09	5.90	0.06	21	1575.0	0.76	445	0.93	15.62	0.00	0.06
8			439	12.50	163.43	5.00	0.07	21			441	1.06	14.24	0.00	0.07
9	2579.0	1.47	502	0.42	12.25	0.14	0.03	22	1960.0	1.02	475	5.09	32.07	3.01	0.14
9			504	0.27	7.43	1.89	0.04	22			473	5.55	40.00	4.62	0.12
10	1145.0	0.91	449	1.93	22.79	2.15	0.08	23	1918.0	0.99	474	0.80	1.77	1.61	0.31
10			446	1.80	22.34	2.34	0.07	23			473	1.00	2.83	2.00	0.26
11	1418.0	0.80	447	8.40	79.40	12.40	0.10	24	2609-2611	1.17	467	5.53	80.00	2.85	0.06
11			453	7.75	70.40	10.81	0.10	24			469	7.01	97.54	3.50	0.07
12	1394.0	0.75	433	3.44	81.14	7.37	0.04	25	2581-2583	1.15	463	6.03	119.81	4.71	0.05
12			434	3.33	80.17	7.71	0.04	25			466	6.34	115.57	1.73	0.05
13	1154.0	0.80	449	0.61	17.07	3.38	0.03	26	2663-2666	1.17	469	4.91	119.32	3.72	0.04
13			449	0.46	15.23	4.46	0.03	26			469	4.92	124.92	3.33	0.04

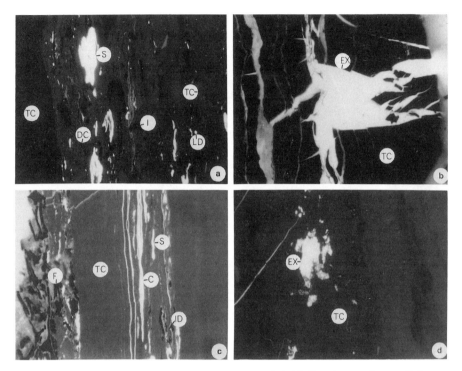

Figure 2. Photomicrographs of Mannville coals initially taken under u.v light (fluorescence) with a water immersion objective lens. Filters: excitation 450-490 nm, beam splitter 510 nm and barrier 520 nm. Long axis of each photomicrograph is 230 micron.

(a) Non-fluorescing telocollinite (TC_1), fluorescing telocollinite (TC_2) and fluorescing desmocollinite (DC), sporinite (S), liptodetrinite (LD) and angular inertinite (I); Simonette, 2196-2200 m.

(b) Exsudatinite (EX) infilling cracks and fractures in non-fluorescing telocollinite (TC); Simonette, 2216-2218 m.

(c) Fluorescing telocollinite bands (TC) associated with cutinite (C), sporinite (S), inertodetrinite (ID) and broken fusinite (F); Pembina 1652.2 m.

(d) Granular exsudatinite (Ex) in a matrix of fluorescing telocollinite (TC); Pembina, 1653.1 m.

Table II. Maceral Composition (Under UV Light) of the Mannville Coals

Sample No.	TC	DC	E and L	I
1	25	60	5	10
2	20	66	4	10
3	30	55	13	2
4	40	58	2	0
5	30	32	8	30
6	20	35	11	34
7	25	35	6	34
8	10	15	5	70
9	57	40	0	3
10	50	37	1	12
11	85	12	3	0
12	20	63	6	11
13	40	36	7	17
14	42	43	3	12
15	40	50	5	5
16	38	46	6	10
17	30	45	10	15
18	23	63	6	9
19	40	38	5	17
20	33	50	4	13
21	35	40	7	18
22	45	42	3	10
23	37	50	1	12
24	44	45	1	10
25	37	47	1	15
26	42	44	1	13

TC - Telocollinite
DC - Desmocollinite
E and L - Exinite and Liptinite
I - Inertinite

anisotropic inertinite as well as pyrolytic carbon, showing a rosette morphology and cross-extinction pattern are also present. Coals examined from wells GDP et al PEM (Ro, max is 0.82% at 1806 m), from Gunn (Ro, max is 0.72% at 1245 m) and from Imperial Isle (Ro, max is 0.78% at 1345 m) (Table I) are dominated by vitrinite with minor liptinite and inertinite (Table II). Round bitumen particles and bitumen staining are present in the last two samples. A coal from the Ermineskin well contains anisotropic inertodetrinite, micrinite, desmocollinite (Table II) and has Ro, max of 0.79% at 1340 m (Table I). One of the shallowest coals in this study, present in the Fairydell well (865 m) has Ro, max of 0.65% (Table I), contains pyrolytic carbon and anisotropic inertodetrinite (Table II). Coals from the Bashaw (Ro, max is 0.76% at 1575 m), Cherhill (Ro, max is 0.77% at 1209 m) and Cherhill (Ro, max is 0.70% at 1310 m) (Table I) are dominated by both telocollinite and desmocollinite with minor sporinite, resinite, vitrodetrinite, macrinite, sclerotinite, semifusinite and inertodetrinite (Table II). Two coals from the Minhik well have Ro, max of 0.99% at 1918 m and 1.02% at 1960 m (Table I). Both are dominated by vitrinite with minor liptinite and inertinite (Table II).

A different situation exists in the Conoco Colt well, where the Mannville coals are much deeper (2582.8-2665.8 m) (Table I). Vitrinite has attained a much higher maturity (Ro, max is 1.15 to 1.17%). The coals are dominated by vitrinite and inertinite, some of the inertinite fragments showing 'strain' anisotropy and extremely anisotropic pyrolytic carbon (Table II). Traces of liptinite, in the form of resinite and broken cutinite were identified in the microbrecciated coal matrix. The rank increase is due to the greater burial depth of the coals and to the effect of structural deformation, which is responsible for the formation of pyrolytic carbon and the microbrecciation.

Rock-Eval Pyrolysis

One of the most obvious characteristics of the Mannville Group coals is the high hydrogen index (HI) and low oxygen index (OI) values (Table I). HI values have a range from 30 to 444 mg HC/g TOC, with numerous samples being in the 200 to 400 range. The OI values are uniformly low (less than 60 mg CO_2/g TOC) but generally less than about 20. This is unusual for coals because most coals have HI less than 300 and OI more than 50, an indication that these coals are enriched in hydrogen and depleted in oxygen compared to humic coals. The production index (PI = $S_1/(S_1+S_2)$) values are generally low (less than 0.10) although two coals have PI values between 0.12 and 0.31. The S_2/S_3 values are generally greater than 5, which is the norm for most coals. The Tmax values are consistent (433-449°C) for the majority of samples, although values as high as 504°C were recorded, reflecting the higher level of maturity (1.47% Ro, max) of the sample. The coals plot along the kerogen Types I and II evolutionary pathway on the pseudo-van Krevelen diagram (Figure 3), thus possibly overestimating the liquid hydrocarbon potential for these types of coals. The anomalously low OI values of the coals could be an artifact of the detection mode because the coals, being in the mature stage of hydrocarbon generation (Ro greater than 0.6%) may have released oxygen as carbon dioxide. The HI vs. Tmax plot (Figure 4) shows that all coals, with the exception of one, are within the 'oil window' and that the composition is an admixture of kerogen Types II and III, in closer

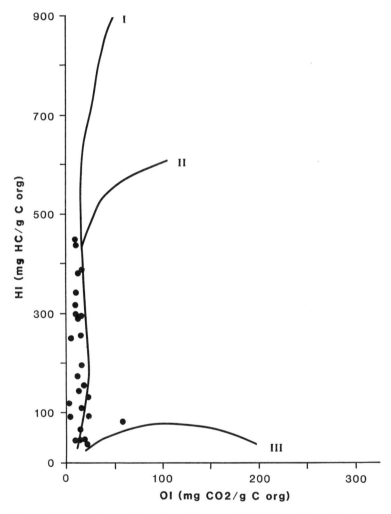

Figure 3. Pseudo-van Krevelen diagram of the Mannville coals studied.

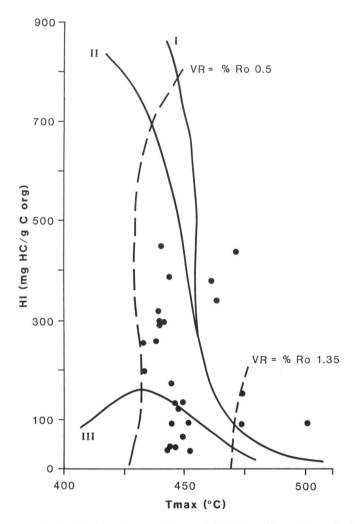

Figure 4. A plot of HI versus Tmax of the Mannville coals studied.

agreement with petrological observations. Such a plot is thus recommended in order to differentiate kerogen evolution pathways and avoid problems with OI.

The relationship between Tmax and %Ro,max of the Mannville Group coals is shown in Figure 5, superimposed on that published by Teichmuller and Durand (1) for world coals. The Mannville coals generally have a slightly higher Tmax when compared to coals having a similar Ro value in the Teichmuller and Durand study. For example, in the Ro, max range of 0.7 to 0.8%, the Mannville coals have Tmax of 435 to 455°C, whereas other world coals have Tmax ranging from 434 to 443°C. It is believed that the Tmax of the Mannville coals does not represent the true level of maturity. One of the least mature coals of this study (Ro, max = 0.70%) has Tmax of 433°C, an intermediate maturity coal (Ro, max = 0.99%) has Tmax of 474°C and the most mature coal (Ro, max = 1.47%) has Tmax equal to 504°C (Table I). The agreement between Rock-Eval Tmax and Ro of vitrinite in the coals is not good in the upper end of the 'oil window' (1.0-1.4% Ro), where Tmax values (465-475°C) translate to Ro of 1.35-1.5%, yet the measured Ro is only 0.99-1.17% (Table I). The agreement is better in the range between onset to 'oil window' and peak of oil generation for Type III kerogen (i.e. coal) and Tmax values of 435-445°C, which correspond to Ro of 0.6-0.8%. This discrepancy between measured vitrinite reflectance and Tmax could be due to a number of factors, but two are the most important in our opinion, which will be discussed later. The Mannville coals have a range of S_1 (free hydrocarbons) from 0.18 to 29.38 mg HC/g rock, with many values over 5 (Table I), while the coals studied by Teichmuller and Durand (1) have S_1 values from 0.4 to 6.6 mg HC/g rock over the same level of maturity. This means that the Mannville coals have a higher content of 'free hydrocarbons' than other world coals. The values of the S_2 hydrocarbons (to be distilled upon heating) range from about 2 to 211.50 mg HC/g rock (Table I) with many values being greater than 20. With such high S_1 and S_2 values in particular, as well as S_2/S_3 ratio greater than 5 in most samples, the samples having HI over 300 should be considered as having a very good potential for liquid hydrocarbon generation. The relationship between vitrinite Ro, max and S_1+S_2 shown in Figure 6 indicates that the highest total potential for hydrocarbon generation is at Ro levels of 0.65-0.75%, which is at the onset of the 'oil window' for Type III organic matter. As thermal maturity increases and hydrocarbons are generated, total potential decreases. This is exactly what is observed with the Mannville coals, with a few exceptions. One group of coals still has some moderate potential even at Ro, max of almost 1.2%, whereas another group has almost no potential at Ro, max of 0.65-0.75%. The production index values are less than 0.1, with the exception of two samples, which is consistent with thermal maturity inferred from vitrinite reflectance. One coal has PI of 0.3 (corresponding to Ro, max of 1.2%), yet the measured Ro, max is only 0.99%. This points to the inappropriateness of using PI only (not the other Rock-Eval indices) for thermal maturity purposes in some cases. The low PI values indicate that volatile or free (S_1) hydrocarbons, normally present in the nC_{10-25} range are absent in the coals.

Discussion

The high hydrogen and low oxygen content of the Mannville Group coals points to initial deposition of the organic matter in an environment where anaerobic conditions

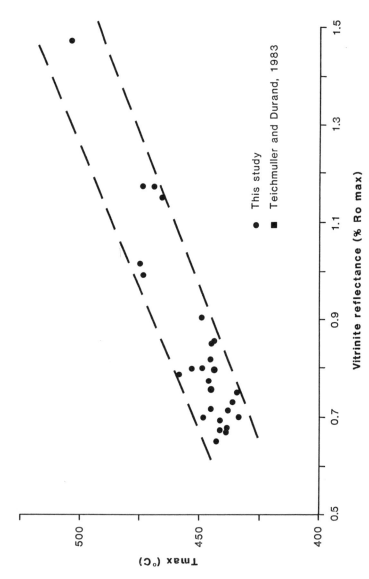

Figure 5. The relationship between vitrinite Ro, max and Tmax of the Mannville coals, superimposed on data from Teichmuller and Durand (2).

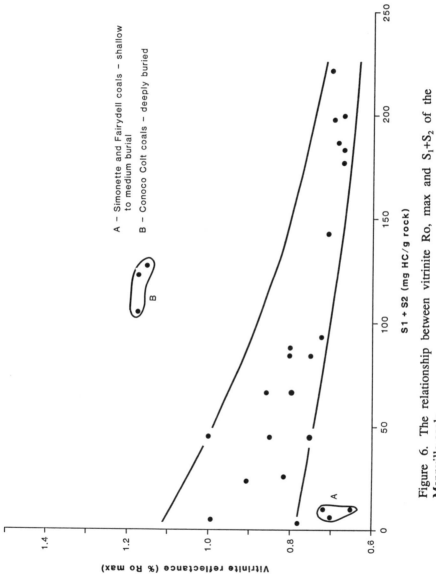

Figure 6. The relationship between vitrinite R_o, max and $S_1 + S_2$ of the Mannville coals.

prevailed and organically-bound oxygen was removed by anaerobic microbes. This resulted in an enrichment of hydrogen in vitrinite; the enrichment in hydrogen in vitrinite tends to suppress its reflectance. It is also known that the fluorescence intensity of vitrinite group macerals is strongly affected by the depositional environment of the coal, particularly the degree of marine influence (18). Any coal formed in an environment in proximity to marine or brackish waters (i.e. lower delta plain, coastal swamp), where the original peat interacted with marine waters during a transgression or a periodic inundation, would have a 'marine influence' imprinted in its vitrinite chemistry and other petrological characteristics. The dependence of vitrinite fluorescence on the coal's depositional environment has been shown by Rathbone and Davis (18) using statistical analysis. They concluded that the marine-influenced vitrinites generally have higher fluorescence intensities. A different approach would be to compare the relative fluorescence intensity, determined qualitatively in this case, to a property of coal known to be influenced by its depositional environment, such as its sulphur or, even better, its boron (B) content.

The studies of Banerjee and Goodarzi (14) and Goodarzi and van der Flier-Keller (15) show that the B content of coals formed under brackish waters is considerably higher than those formed under fresh waters and that there is a clear correlation between water chemistry and B content of coal. Rathbone and Davis (18) found very good relationship between total sulphur as well as pyritic sulphur with the fluorescence intensity of vitrinite in the Carboniferous Lower Kittaning coals. The total sulphur content of a suite of 28 Mannville coals, from the southern plains of Alberta, is up to 23.46% (24), mostly pyritic. The environmental conditions which favoured pyrite formation also may have been responsible for the brighter fluorescence intensity of telocollinite and desmocollinite in these coals. Also, the B content is as high as 1144 ppm, indicative of a very strong brackish-water influence on the vitrinite chemistry and fluorescence intensity. Normally, freshwater coals have B content of less than 50 ppm, slightly-brackish coals between 50 and 110 ppm and brackish coal greater than 110 ppm (14). The Mannville coals have some of the highest B content ever determined in coal, which points to a possible relationship between coal depositional environment and fluorescence of vitrinite.

According to one theory, the cause of the secondary fluorescence of vitrinite in coal (i.e. after the first coalification 'jump' at more than 0.50% Ro) is due to the generation of liquid bituminous substances from the liptinite and exinite group macerals due to increased thermal maturity and subsequent absorption by the vitrinite macerals, especially at areas in close proximity to hydrogen-rich macerals. Although the exinite and liptinite group macerals volume in the Mannville coals is not high (less than 20%), the generation of hydrocarbons is evident by the formation of granular micrinite and of exsudatinite, the latter in both massive and granular form (Figure 2b and 2d). The macerals sporinite, cutinite, suberinite, resinite and alginite in these coals may also have contributed to hydrocarbon generation, especially if favourable conditions (i.e. dysaerobic putrefaction/fermentation, compaction and homogenization of biopolymers) existed (19). Exsudatinite occurs as crack- and cavity-filling, indicating that it once was mobile, oozing out of hydrogen-rich macerals and into the cracks (Figure 2b).

Regional thermal maturity patterns in the Alberta Plains indicate that the Ro of the Mannville Group coals should be higher by 0.1 to 0.3% (Pearson, D.E., personal communication, 1988). Thus, the vitrinite Ro suppression may be attributed to at least two factors. They are: 1) the hydrogen-rich nature of the vitrinite in the coals, a reflection of the brackish water nature which dominated during peat accumulation and subsequent coal formation; and 2) the generation and absorption of hydrogen-rich secondary bitumen in the vitrinite structure. Mukhopadhyay (8) observed that mixinitic (>20% desmocollinite/telocollinite, 5% saprovitrinite, 20-25% liptinite, <15% inertinite) coals and sapropelic coals are generally perhydrous and contain abundant aliphatic carbon compared to humic coals. Seven of the twenty six Mannville coals analyzed have HI values between 225 and 350 mg HC/g Corg and four have HI above 350 (Table I). Under hydrous pyrolysis, coals of similar petrographic composition, depositional environment, rank and HI values have the ability to generate liquid hydrocarbons below 1.0% Ro (11). Lin and Davis (20) state that it is unlikely that the second factor, listed above, is the primary cause of the secondary fluorescence of vitrinite because saturated aliphatic hydrocarbons normally do not fluoresce. However, if the vitrinite matrix if fine-grained (desmocollinite, saprovitrinite), any impregnation of bitumen, generated during thermal maturation, could be responsible for suppression in vitrinite Ro (8). The difference in the degree of fluorescence intensity between telocollinite and desmocollinite, is evident in Figure 2a. Desmocollinite, in the middle of the photomicrograph, being intimately associated with exinite group macerals is more prone to bitumen incorporation than the thick-banded telocollinite, to the left of the same photo. An interesting observation is that, despite the high level of thermal maturity indicated by measured Ro, max in a number of samples (1.15-1.17%) (Table I), which places them in the late catagenetic stage of hydrocarbon generation, the HI values for a number of Mannville coals are still greater than 300 mgHC/g Corg. This implies that the coals were most likely much richer in hydrogen than normal coals at lower levels of maturity prior to hydrocarbon generation. However, the use of HI by itself as an indicator of the hydrocarbon-generating potential of the coals should be treated with caution (21). For example, samples 24,25 and 26 have %Ro, max of 1.15-1.17 (toward the end of the oil window), yet their HI values are among the highest (332-439)(Table I). On the other hand, samples 4,5 and 6 have %Ro, max of only 0.67 to 0.70 but still similar HI values (319-444) (Table I). This indicates the lack of a good correlation between the two parameters in these unusual coals only.

A marine-influenced coal deposition usually results in the incorporation of bacteria-derived lipids into the vitrinite structure at the early stages of diagenesis (18). The occurrence of bacteria-derived products results in a contribution to vitrinite fluorescence even at high levels of maturity. Two of these products are steranes and terpanes, which are detected in coals and sediments using gas chromatography-mass spectrometry (GC-MS) by selectively monitoring their fragmentograms, such as the $m/z = 217$ for steranes and $m/z = 191$ for terpanes. The GC-MS analyses of the steranes in the solvent extract of a number of Mannville coals near the vicinity of the present study area, reported in Goodarzi et al. (in press), show the predominance of C_{29} steranes with both S and R isomers present in the $m/z = 217$ ion. Less abundant in the samples are C_{29} diasteranes. The $m/z = 218$ fragmentogram is also dominated by C_{29} (S and R) steranes although small amounts of C_{27} and C_{28} steranes are also present. Hopanes (pentacyclic triterpenoid compounds) and C_{31} and C_{32} homohopanes were also detected. The dominance of C_{29} steranes over C_{27} and C_{28} in both the m/z

= 217 and m/z = 218 traces indicates contribution from higher plants with a very small algal or bacterial contribution (22). This particular feature indicates that any hydrocarbons present in these Mannville coals were most likely derived from the organic matter within the coal, with very little allochthonous contribution. The ratio of the C_{29} S to C_{29} R steranes is almost 1:1, which points to a mature source (0.8-0.9% Ro, max), in agreement with the vitrinite reflectance data of the Mannville coals containing these biomarkers.

The ratio of C_{27}:C_{28}:C_{29} steranes of Upper Cretaceous oils in western Canada sedimentary basin is 1:1:1 (Allan, J., personal communication, 1993), considerably different when compared to the ratio of the same compounds in the Lower Cretaceous Mannville coal extracts where the C_{29} steranes dominate. The fluorescence and greasy or oily appearance of the Pembina well core samples could be interpreted as either hydrocarbons produced *in situ* by the organic matter in the Mannville coals due to increased thermal maturation or not *in situ*-formed hydrocarbons, but absorbed by the coaly matrix en route to a reservoir. Oils in the same basin, sourced from sub-Cretaceous strata (Mesozoic or Paleozoic) do not have the 1:1:1 ratio of terpanes seen in Upper Cretaceous oils. Therefore, the possibility of some allochthonous contribution cannot be ruled out. Based on the data of this study, it is most likely that the coals themselves have generated at least some of the hydrocarbons and subsequently absorbed them, thus resulting in the vitrinite reflectance suppression. In other words, these Mannville coals may have acted both as source and as reservoir for the hydrocarbons. Although coal is viewed mainly as a gas producer because of its organic matter type, there are proven examples of coal seams being source rocks (23). The problem appears to be the inefficiency of coal seams to expel any generated hydrocarbons (unlike organic-rich shales) due to the coal's macro- and microporosity and permeability. As a result, coal acts as an organic 'sponge', especially at the subbituminous and high-volatile rank levels. For thin coals of higher rank, such as some of the Mannville coals of this study, oil saturation could have taken place in the network of desmocollinite and liptodetrinite. This network, which excludes exsudatinite in telocollinite, acts as a primary migration pathway for liquid hydrocarbons. Any hydrocarbons trapped as a 'mobile phase' would remain in the coal's matrix and cracked to gas at higher levels of maturation, as proposed by Mukhopadhyay (8).

Conclusion

Some of the Mannville coals in the Alberta Plains are highly enriched in hydrogen when compared to most world coals, even at high levels of thermal maturity (~1.2 %Ro, max). This points to the possibility that the original coals were probably much enriched at the onset of the 'oil window'. The fluorescence intensity of both telo- and desmocollinite in the coals is related to the marine influence during peat deposition, which is also indicated by the high total sulphur and extremely high boron contents of the coals. Maximum reflectance of vitrinite in the Mannville coals has been suppressed by as much as 0.3%, and this is believed to be not due to liptinite abundance but rather due to the chemistry of vitrinite, as a result of depositional environment, and to a certain degree absorption of bitumen by the vitrinite matrix. Although organic geochemistry shows that at least some of this bitumen may be derived *in situ* by the coals themselves, the possibility of the coals having absorbed *in situ*-derived bitumen cannot be ruled out.

Acknowledgments

We would like to thank Dr. P.K. Mukhopadhyay of Global Geoenergy Research Ltd., Halifax, Canada, for an invitation to contribute toward this volume and the reviewers for their helpful comments. A special thanks to Mr. Glenn McWilliams, Norcen Resources Ltd., Calgary, Alberta for allowing publication of the results and Mrs. Rita Schultz, Alberta Research Council for assistance with word processing of the manuscript. Alberta Research Council Contribution No. 2155.

Literature Cited

1. Teichmuller, M. and Durand, B. *Fluorescence Microscopical Rank Studies on Liptinites and Vitrinites in Peat and Coals, and Comparison with Results of the Rock-Eval Pyrolysis;* Int. J. Coal Geol., 1983; Vol. 2, pp 197-230.
2. Jones, R.W. and Edison, T.A. Microscopic Observations of Kerogen Related to Geochemical Parameters with Emphasis on Thermal Maturation, In *Low Temperature Metamorphism of Kerogen and Clay Minerals;* Oltz, D.F., (Ed.): Symposium in Geochemistry, Pacific Section, Soc. Econ. Paleont. Mineral., LA, CA, USA, 1979; pp 1-12.
3. Price, L.C. and Barker, C.E. *Suppression of Vitrinite Reflectance in Amorphous Rich Kerogen - A Major Unrecognized Problem;* J. Pet. Geol., 1985, Vol. 8, pp 59-84.
4. Hutton, A.C. and Cook, A.C. *Influence of Alginite on the Reflectance of Vitrinite from Joadja, NSW, and Some Other Coals and Oil Shales Containing Alginite;* Fuel, 1980; Vol. 59, pp 711-714.
5. Kalkreuth, W.D. *Rank and Petrographic Composition of Selected Jurassic Lower Cretaceous Coals of British Columbia;* Canadian Pet. Geol. Bull., 1982; Vol. 30, pp 112-139.
6. Snowdon, L.R., Brooks, P.W. and Goodarzi, F. *Chemical and Petrological Properties of Some Liptinite-rich Coals from British Columbia;* Fuel, 1986; Vol. 65, pp 460-472.
7. Goodarzi, F., Davies, G.R., Nassichuk, W.W. and Snowdon, L.R. *Organic Petrology and Rock-Eval Pyrolysis of the Lower Carboniferous Emma Fiord Formation in Sverdrup Basin:* Canadian Arctic Archipelago, Mar. Pet. Geol., 1987; Vol. 4, pp 132-145.
8. Mukhopadhyay, P.K. *Hydrocarbon Generation from Deltaic and Intermontane Fluviodeltaic Coal and Coaly Shale from the Tertiary of Texas and Carboniferous of Nova Scotia;* Org. Geochem. M, 1992; Vol. 17, pp 765-783.
9. Goodarzi, F., Gentzis, T., Snowdon, L.R., Bustin, R.M., Feinstein, S. and Labonte, M. *Effect of Mineral Matrix and Seam Thickness of Vitrinite in High to Low Volatile Bituminous Coals: An Enigma;* Mar. Pet. Geol., 1993; Vol. 10, pp 162-171.

10. Teichmuller, M., *Recent Advances in Coalification Studies and Their Application to Geology, in Coal and Coal-bearing Strata - Recent Advances;* Scott, A.C.,(Ed.), Geol. Soc. London Spec. Publ., 1986; Vol. 32, pp 127-169.

11. Goodarzi, F., Snowdon, L.R., Gentzis, T. and Pearson, D.E., In Press *Petrological and Chemical Characteristics of Liptinite-rich Coals from Alberta, Canada;* Mar. Pet. Geol.

12. Hacquebard, P.E. *Rank of Coal as an Index of Organic Metamorphism for Oil and Gas in Alberta, Chapter 3, The Origin and Migration of Petroleum in the Western Canadian Sedimentary Basin, Alberta - A Geochemical and Thermal Maturation Study;* Deroo et al. (Eds.), Geol. Sur. Canada, Bull. 262, 1977; pp 11-22.

13. Osadetz, K.G., Pearson, D.E. and Stasiuk, L.D. *Paleogeothermal Gradients and Changes in the Geothermal Field of the Alberta Plains;* In Cur. Res., Pt. D, Geol. Sur. Canada, Pap.90-1D, 1990; pp 165-178.

14. Banerjee, I. and Goodarzi, F. *Paleoenvironmental and Sulphur-boron Contents of the Mannville (Lower Cretaceous) Coals of Southern Alberta, Canada;* Sedim. Geol., 1990; Vol. 67, pp 297-310.

15. Goodarzi, F. and van der Flier-Keller, E. *Organic Petrology and Geochemistry of Intermontane Coals from British Columbia, Canada;* Chem. Geol., 1988; Vol. 7, pp 127-141.

16. *International Committee for Coal Petrology, 1975 International Handbook of Coal Petrology,* 3rd edition 1971, supplement to 3rd edition 1975, Centre National de la Recherche Scientifique, Paris.

17. Espitalie, J.M., Madec, M., Tissot, B., Mennig, J.J. and Leplat, P. *Source Rock Characterization Method for Petroleum Exploration:* Proceedings of the 9th Annual Offshore Technology Conference; 1977; Vol. 3, pp 439-448.

18. Rathbone, R.F. and Davis, A. *The Effects of Depositional Environment on Vitrinite Fluorescence Intensity;* Org. Geochem., 1993; Vol. 20, pp 177-186.

19. Khavari-Khorasani, G. Oil-prone Coals of the Wallon Coal Measures, Surat Basin, Australia; In *Coal and Coal-bearing Strata;* Scott, A.C., (Ed.), Geol. Soc. London, Spec. Publ. 32, 1987; pp 303-310.

20. Lin, R., and Davis, A. *A Fluorogeochemical Model of Coal Macerals;* Org. Geochem., 1988; Vol. 12, pp 363-374.

21. Lewan, M.D. and Williams, J.A. *Evaluation of Petroleum Generation from Resinites by Hydrous Pyrolysis;* Amer. Assoc. Pet. Geol. Bull., 1987; Vol. 71, pp 207-214.

22. Chaffee, A.L., Hoover, D.S., Johns, R.B. and Schweighardt, F.K. Biological Markers Extractable from Coal; In *Biological Markers in the Sedimentary Record;* Johns, R.B., (Ed.), Elsevier, Amsterdam, 1986; pp 311-345.

23. Fowler, M.G., Gentzis, T., Goodarzi, F. and Foscolos, A.E. *The Hydrocarbon Potential of Lignites from Northern Greece as Determined by Organic Petrological and Geochemical Techniques;* Org. Geochem., 1991; Vol. 17, pp 805-826.

RECEIVED June 9, 1994

MOLECULAR CHARACTERIZATION
OF VITRINITE MATURATION

Chapter 8

Coalification Reactions of Vitrinite Derived from Coalified Wood

Transformations to Rank of Bituminous Coal

Patrick G. Hatcher[1], Kurt A. Wenzel[1], and George D. Cody[2]

[1]Fuel Science Program, 209 Academic Projects Building,
Pennsylvania State University, University Park, PA 16802
[2]Chemistry Division, Argonne National Laboratory,
9700 South Cass Avenue, Argonne, IL 60439

Vitrinite, a major component of most coals, is derived from degraded wood in ancient peat swamps. Organic geochemical studies conducted on a series of coalified wood samples derived mostly from gymnosperms have allowed the development of a chemical reaction series to characterize the major coalification reactions which lignin, the major coal-producing component of wood, undergoes. These involve mostly hydrolysis of aryl ethers, reduction of alkoxyl side chains, alkylation of reactive aromatic sites, and finally condensation reactions of phenolic moieties. Thus, lignin is transformed to catechol-like structures and these are converted to phenol-like structures. The former reactions are well understood to occur during peatification by the action of microorganisms, but the latter have not been shown to be possible in other than model systems. To demonstrate that the catechol-like structures can be made to evolve to phenol-like structures under natural conditions, we attempted to simulate natural burial processes by artificially coalifying a sample containing mostly catechol-like structures under conditions of hydrous pyrolysis. These conditions did indeed induce transformations of catechol-like structures to phenol-like structures, but the reactions do not reproduce well the coalification reactions with regard to aliphatic structures in coal.

It is well recognized that coal is a complex assemblage of macromolecularly discreet components of varied origin. Accordingly one would expect the chemistry of each type of discreet component to be different, though some may exhibit certain similarities. Defining the chemical structural compositions of all these discreet components is a formidable task, even by today's standards of rapid sophisticated analyses. Perhaps one aspect of coal structure which has prevented more thorough investigations of the components is the fact that many components, macerals in the vernacular, exist as minute entities finely comminuted within a matrix composed of other macerals. The inability to cleanly separate macerals for detailed chemical analysis has placed some severe constraints on our knowledge of their chemical structural compositions. Separation methods based on density have allowed some progress in this area (1,2), but these methods cannot provide reliable separations for all macerals.The recent introduction of a new technique of laser pyrolysis (3,4) which has

0097–6156/94/0570–0112$08.72/0
© 1994 American Chemical Society

the capability of focusing on the pyrolytic chemistry of macerals within a laser beam's path of 25 µm, holds some promise for future work. Thus at present, we are forced to rely on hand-picking methods to reliably isolate macerals for chemical analysis. The major drawback here is the inability to hand-pick all but the macerals existing in large physical domains. Even in such cases, we can never be certain of maceral purity. Our only hope of defining coal's chemical structural composition is by a systematic isolation and characterization of the major maceral types, as a start, and then the development of new techniques of microanalysis. In cases where macerals cannot be cleanly isolated at the macroscopic scale, we must resign ourselves to separation by density gradient centrifugation of finely ground coal. However, in cases where macerals can be isolated on a macroscopic scale, it is best to resort to hand-picking techniques for isolation. Accordingly, the major maceral type which can be easily recognized and physically isolated is vitrinite derived from coalified wood specimens.

The structural evolution of vitrinite has been previously studied in our group through detailed characterization of coalified wood which spans the entire coalification range (*5-7*). The use of coalified logs was demonstrated to be especially useful as it removes the superposed problem of maceral scale heterogenieties from the problem of identifying fundamental transformations which delineate the coal's chemical structural evolution. The net result of these detailed studies using ^{13}C NMR and pyrolysis/gas chromatography/ mass spectrometric (py/gc/ms) methods is the identification of several key chemical structural transformations which typify the coalification of woody material into high volatile bituminous coal.

Although, the coalification series represents a continuum of parallel and serial processes; several principle stages of coalification are clearly evident. The initial biochemical stage of coalification is characterized by a complete loss in hemicellulose, a significant reduction in cellulose, and selective preservation, with minimal alteration, of lignin derived material (*8,9*). Early diagenetic changes accompanying the transformation from brown coal into lignite result in a near complete removal of cellulose and some modification of the lignin, a macromolecular material composed of methoxyphenols with a polyhydroxypropanol side-chain as basic building blocks. These modifications are, dominantly, rearrangement of the alkyl-aryl ether bond that links the methoxyphenols together to yield methoxyphenols linked by an aryl-alkyl bond between the structural units, as well as some demethylation of methoxy groups yielding catechol-like structures (*7*). The transformation of coalified wood from lignite through the subbituminous rank range to high volatile C rank bituminous coal is characterized by transformations which result in complete demethylation of methoxyphenols to catechols and a subsequent reduction of the catechol-like structures, presumably through reaction, to form phenol-like structures. The focus of this paper is to review our current understanding of the evidence for these aforementioned chemical transformations and to present some new data relating to the specific reactions responsible for the major part of oxygen loss during coalification.

Recently, Siskin and Katritzky (*10*) have demonstrated that many of the reactions which typify coalification are facilitated, if not completely initiated, by the presence of water. They recognized that, since coalification occurs in a water saturated system, water may play an integral role in the chemistry of coalification. From their study of an enormous number of reactions with model compounds, several mechanisms with direct bearing on the chemical structural evolution of coal are clear. The demethylation of methoxybenzene, for example, has been shown to be acid catalyzed yielding phenol as the predominant product (*11*). The mechanism by which the alkyl-aryl ether linkage (known as the ß-0-4 linkage) of the modified lignin is rearranged to the ß-C-5 linkage as proposed by Hatcher (*7*) to explain coalification of lignin in wood to brown coal and lignite may reasonably be expected to parallel the mechanism that Siskin et al.(*12*) demonstrated for their model compound study of the hydrous pyrolysis of benzylphenylether. They observed a significant yield of 2 benzyl phenol; this product is a perfect analog for the rearrangement necessary to yield the ß-C-5 linkage found to be important in brown coal wood.

Currently, there is no mechanism proposed for the conversion of catechol-like structures to phenol-like structures during coalification. However, this transformation appears to be a principal means by which the oxygen content of coal is reduced during the transformation from brown coal and lignitic wood through wood coalified to the subbituminous range (7) and is, therefore, of considerable interest. Although Siskin and Katritzky (10) have not investigated the chemistry of catechols specifically, they did study 4-phenoxyphenol. It appears reasonable to assume that the reactions involving condensed catechols, as might be found in coal, may follow a similar reaction chemistry. The present paper sets out to investigate this possibility as well as to investigate the role of hydrous pyrolysis on coalification in a broader sense by initiating artificial coalification studies on a lignitic coalified log. By comparing the chemical structural evolution of the lignitic log, which has been subjected to hydrous pyrolysis, to the results from previous studies of natural coalification, it should be possible to derive the most probable mechanism for the catechol reduction pathway. In addition to this specific transformation, the success of artificial coalification in general is considered through the comparison between the residues of treated lignitic log with essentially equivalent rank, naturally coalified, logs.

Review of the Coalification of Wood

The advent of some new techniques in the 1980's has provided the methodology for establishing a better understanding of the chemistry of coal and its maceral components. The application of these techniques, principally ^{13}C nuclear magnetic resonance (NMR) and flash pyrolysis/gas chromatography/mass spectrometry (py/gc/ms), to studies of series of coalified wood samples has provided a new glimpse of the processes responsible for coalification. By comparisons of the chemical compositions of wood at various stages of coalification, one can infer specific reactions responsible for coalification.

Peatification. It was made especially clear from studies of fresh wood and peatified wood (8,9,13-15) that cellulosic components of wood, the ones contributing more than 60% of the structure, essentially are mineralized or degraded and lost within a short span of time geologically. Thus, the cellulosic components do not play a significant role in the structural make-up of coalified wood. The lignin, however, is selectively preserved in a relatively unaltered state during peatification and is the substance which eventually forms the coalified wood's vitrinitic component.

Perhaps the most astounding aspect of this degradative process is the selectivity and structural precision with which it occurs. Essentially all of the major mass component of wood, the cellulose, is lost. One might infer that tremendous physical destruction ensues; however, peatified wood remarkably retains its morphology. This can clearly be seen in Figure 1 which shows an SEM photomicrograph of degraded wood in peat. Delicate wood structures such as bordered pits and cell wall tracheids appear physically intact even though all the cellulose has been degraded. Apparently, the wood has been degraded by bacteria which use extracellular enzymes to enable destruction of the cellulosic materials. This process apparently does not involve physical maceration.

Coalification to Brown Coal and Lignite. The lignin which survives this initial degradative process relatively unscathed, is eventually altered over the course of geological time as wood in peat is buried in sedimentary systems. We know this from studies of coalified wood in brown coals and lignites (5-7,13-17). By comparing the chemistries of peatified wood, mainly lignin, with brown coal or lignitic woods of the same family (e.g., gymnosperms, angiosperms), we can decipher reactions which

Figure 1. SEM photmicrograph of peatified white cedar wood from the Dismal Swamp, VA. Bordered pits and cell walls appear well preserved even though nearly all the cellulose has been degraded and lost from the wood.

might be responsible for observed changes. We certainly can identify the nature of chemical changes which occurred. These are depicted in Figure 2.

Perhaps the most chemically and microbiologically labile bond in lignin is the bond linking the methoxyphenolic aromatic monomers together, the β-O-4 bond. Evidence that this bond is cleaved during coalification was provided by Hatcher (7), from an examination of the NMR spectra of brown coal woods. If we consider that all such bonds in lignin are broken and the fact that approximately 60% of the bonds in lignin are of this type, then we might expect the lignin, or the peatified wood, to be completely macerated because rupture of this bond would release molecular fragments which are likely to be soluble in water. Examination of SEM photomicrographs of brown coal wood in which such bond ruptures have occurred shows that physical disruption of the wood anatomy does not occur to any great extent (Figure 3). Delicate structures such as bordered pits would not physically survive a maceration. Also, the SEM photos do not provide any evidence for dissolution or even partial dissolution of cell walls. This must imply that another reaction maintains the macromolecular integrity of the lignin.

The reaction most likely responsible for maintaining structural integrity must be one which allows connectivity between methoxyphenolic structures in lignin. Thus, we must maintain some bond between the aromatic units. Alkylation following β-O-4 bond rupture is the likely reaction. Botto (18) has shown that such an alkylation is possible when lignin labeled at the β carbon is subjected to artificial coalification in the presence of clays. In essence, the rupture of the β-O-4 bond releases a carbocation, the β carbon on the sidechain, which is a good electrophile and will attack positions on the aromatic ring of adjacent phenolic structures which are susceptible to electrophilic substitution (e.g., the C-5 predominantly). The most likely aromatic ring to be alkylated is the one from which the β-O-4 bond rupture occurred. The rupture of the β-O-4 bond would produce a phenol which would likely activate the C-5 site to alkylation. Evidence that such a reaction occurs in brown coal wood comes from the NMR data which show increased aromatic ring substitution (6,7). Dipolar dephasing NMR studies indicate that brown coal wood samples have fewer protons per ring than their respective counterparts in peat, indicating that, on the average, one of the ring sites has become covalently bound to an atom other than hydrogen. The NMR data also indicate that the additional substituent atom is not an oxygen but a carbon atom. It is clear that the data all point to the fact that cleavage of the β-O-4 bond during coalification of the lignin in peatified wood leads to alkylation of an adjacent ring by the resulting carbocation, and this overall process leads to a maintenance of the physical integrity of the wood.

To maintain this integrity observed unambiguously from SEM data (Figure 3) it is necessary that the above reaction occur without much structural rearrangement of the lignin. In other words, the reaction must proceed rapidly and must occur at a relatively proximal site to the β-O-4 cleavage. If one considers a lignin model as one which is random in the connectivity between methoxyphenolic units (19,20), then the likelyhood that structural order will be maintained with the above-mentioned transformation is minimal. In a random model, distances between aromatic reaction centers are variable and some significant physical disruptions will ensue if the carbocations formed from β-O-4 bond rupture have to link up with aromatic centers which are more than just local.

Faulon and Hatcher (21) have recently proposed a new model for lignin which would overcome this problem. The model is one which is ordered rather than random. The order is believed to derive from the fact that the methoxyphenolic units linked by

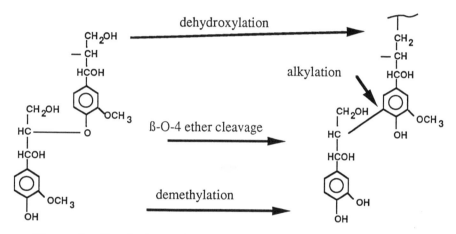

Figure 2. Chemical reactions proposed for the transformation of gymnospermous lignin in peatified wood to brown coal and lignitic gymnospermous wood.

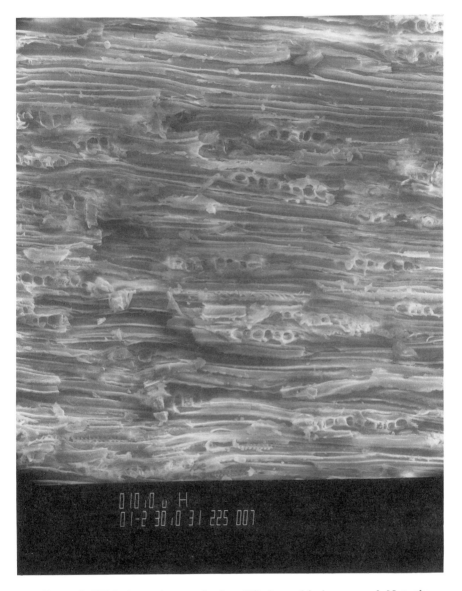

Figure 3. SEM photomicrograph of coalified wood in brown coal. Note the excellent physical preservation of cell walls and other elements characteristic of wood cells.

the β-O-4 bonds in lignin exist in an helical conformation. With such a configuration, cleavage of this bond releases a carbocation which is in relatively close proximity to the C-5 site on the aromatic ring. The alkylation of this site causes minimum disruption of the helical order as a β-C-5 bond is formed (J.-L. Faulon, personnal communication). Thus, we might also expect minimum disruption of the physical integrity of the sample, consistent with what is observed in brown coal wood.

Another important coalification reaction which is observed to occur through the lignite stage of coalification is the cleavage of other aryl-O bonds in lignin. The specific bonds are those of methoxyl groups attached to the aromatic ring. Both NMR and py/gc/ms data show that this is happening (15,16). The chemical degradative data of Ohta and Venkatesan (17) also show this reaction. All these studies indicate that the loss of methoxyl is through a demethylation process whereby the bond between the methyl carbons and the oxygen attached to the ring is cleaved. The molecular modeling studies show that the helical configuration is not affected by such a reaction (22). The resulting structure of the coal contains a phenol where there once was a methoxyl group, and the structure is now said to have one similar to that of catechols, if one considers that the cleavage of the β-O-4 bond has already occurred.

Additional changes in lignin structure are more subtle, but somewhat evident from NMR data. In lignin and peatified wood, the sidechains are hydroxylated at both the α and γ sites. Loss of hydroxyl groups from the sidechains is another reaction resulting from coalification of the lignin. NMR resonances attributable to these hydroxyls diminish substantially during coalification through to the rank of subbituminous coal. Simple loss of the sidechain units by a pyrolytic process would explain the loss of these resonances, but this would lead to a significant increase in aromaticity. Similarly, loss of hydroxyls probably does not proceed through a simple elimination reaction. Such a reaction would produce olefins as side-chain structures and the carbon resonances for these would overlap aromatic carbon resonances. Because aromaticities of coalified wood samples do not increase over the course of the observed loss of hydroxyls, it is likely that simple reduction of the hydroxyls to alkyl groups is the preferred pathway. This would shift the NMR resonances into the alkyl region of the spectra and preserve carbon aromaticities. Such a reaction would also be consistent with the physical structural data, because reduction of hydroxyls would maintain the presumed helical conformation and would cause minimum disruption of the macromolecular structure. Pyrolytic loss of the sidechain carbons would likely macerate the structures beyond recognition. Clearly, to preserve physical integrity to the rank of brown coal or even lignite, the side chains must not be lost.

Coalification to Subbituminous and Bituminous coal. The transformations of brown coal and lignitic woods to higher rank involve some significant changes in chemical structure. This is perhaps the primary cause for the change in physical morphology often observed as coal becomes more lustrous and the virtrinite becomes more homogeneous.This transfromation has often been referred to as gelification (23). Figure 4 shows the SEM of a sample of wood coalified to the rank of subbituminous coal. The wood cells have become deformed significantly, presumable due to increased burial pressure, and some cells have actually been annealed to a homogeneous mass.

While structures in brown coal and lignite are dominated by the catechol-like rings arranged in an helical conformation perhaps, the major components of subbituminous and bituminous woods are phenol-like structures. Evidence for the transformation of catechol-like structures to phenol-like structures as shown in Figure 5 comes from both NMR and py/gc/ms data (7,24). The NMR data clearly indicate a loss of aryl-O carbon. In lignite, aryl-O carbons account for approximately 2 of the six aromatic carbons on the ring, whereas in subbituminous and bituminous coal, only 1

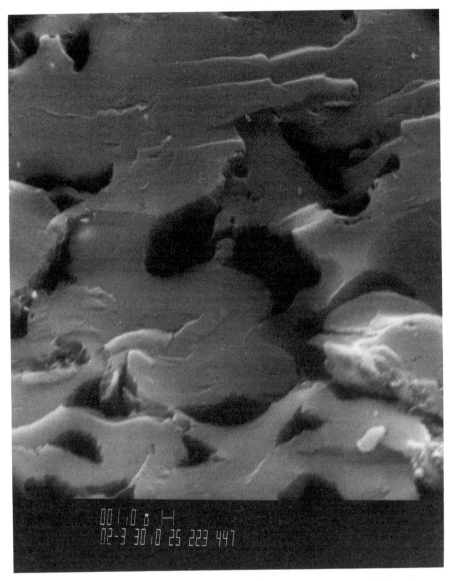

Figure 4. SEM photomicrograph of coalified subbituminous wood showing compressed and deformed cells, some of which appear to be annealed (upper center).

of six aromatic carbons is an aryl-O carbon. Pyrolysis data demonstrate the same observation, with lignitic wood samples being rich in catechols and subbituminous wood samples being rich only in phenols and alkylated phenols. Elemental data show that a significant diminution of oxygen content can explain these tranformations.

Coalification to the rank of high-volatile bituminous coal leads to further reductions in oxygen contents for coalified wood. Considering the fact that the amounts of aryl-O do not change significantly, it is likely that the changes imply a further condensation of phenols to aryl ethers or dibenzofuran-like structures as shown in Figure 6 (*24*). The py/gc/ms data confirm this as more alkylbenzenes and dibenzofurans are observed. The alkylbenzenes in pyrolyzates could arise from thermal cracking of dibenzofuran or diaryl ethers during flash pyrolysis. Also, the increased quantities of condensed aromatic rings, naphthalenes and fluorenes, in pyrolyzates suggests that aromatic ring condensation is occurring. It is likely that this condensation will disrupt the helical structure proposed. At this time we have little to offer in the way of a mechanism for this condensation. Aromaticity of coalified wood appears to increase at this rank (*16*), suggesting that one possible route for the formation of condensed rings is ring closure and aromatization of the alkyl sidechains. Further studies are needed to verify such a pathway.

Perhaps the next most apparent difference between woods coalified to the rank of subbituminous / bituminous coal and those at the rank of lignite or lower rank is the lack of oxygenated alkyl structures (i.e., alkyl hydroxyls or alkyl ethers). The NMR data for subbituminous and bituminous coalified woods show essentially baseline in this region of the spectra, evidence that such functional groups are not significant. Thus what were originally hydroxylated lignin sidechains have been altered. As discussed above, it is most likely that the lignin-derived hydroxyl groups have been reduced rather than lost by pyrolysis of the sidechain. The reduction of hydroxyls is essentially complete at the rank of subbituminous coal.

Consequently, the structural composition of subbituminous coal is that of a lignin structure which has lost its methoxyl groups (via demethylation and dehydroxylation) and all its hydroxyls and alkyl ethers. Presumably, all these reactions can occur with minimal disruption of the three dimensional network inherited from lignin. Indeed, some semblance of cellular morphology still remains at the rank of subbituminous coal (Figure 4). Pressure and temperature begin to combine at these and higher ranks to have a significant effect on physical morphology of wood. Thus, it becomes less clear whether loss of morphology is purely a physical or chemical phenomenon or both. With such a disruption in the macromolecular structure of wood brought on by the formation of condensed ring systems, it seems reasonable that cellular morphology which persists well up to the rank of bituminous coal begins to degrade into a homogeneous glassy appearance with little semblance of cell wall boundaries at ranks of bituminous coal or higher.

To acertain the effect of chemistry on physical morphology, it is imperative that we understand better the nature of the reactions which transform the catechol-like structures of lignitic woods to the phenol- and diarylether-like structures of subbituminous and bituminous woods. Herein lies the focus of this paper.

Experimental

A crushed sample of a coalified gymnospermous wood described previously as the Patapsco lignite (*16*) was heated under hydrous conditions in a 22 mL autoclave reactor described in a previous report (*26*). The reactor was charged under nitrogen with approximately 1 gram of coal and 7 milliliters of deionized, deoxygenated water following a procedure similar to that of Siskin et al. (*12*). The tubing bomb reactor was consecutively pressurized to 1000 psi and depressurized with nitrogen three times to ensure all the oxygen was removed from the reactor. Finally it was depressurized to atmospheric pressure before heating. The bomb was inserted into a heated fluidized

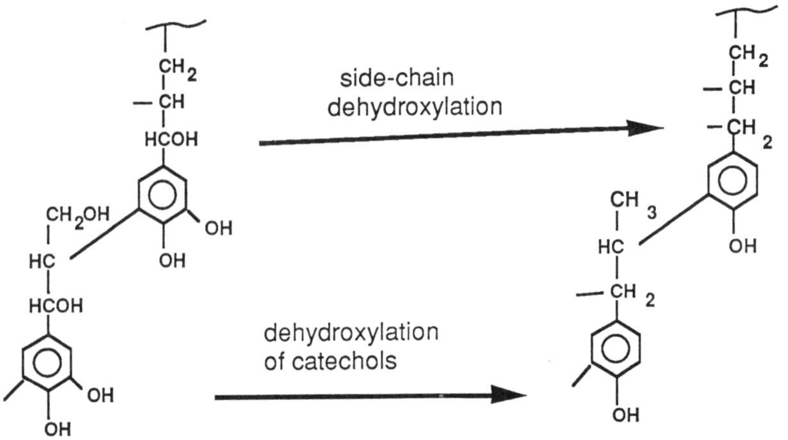

Figure 5. Proposed reactions responsible for the transformation of lignitic gymnospermous wood to subbituminous wood.

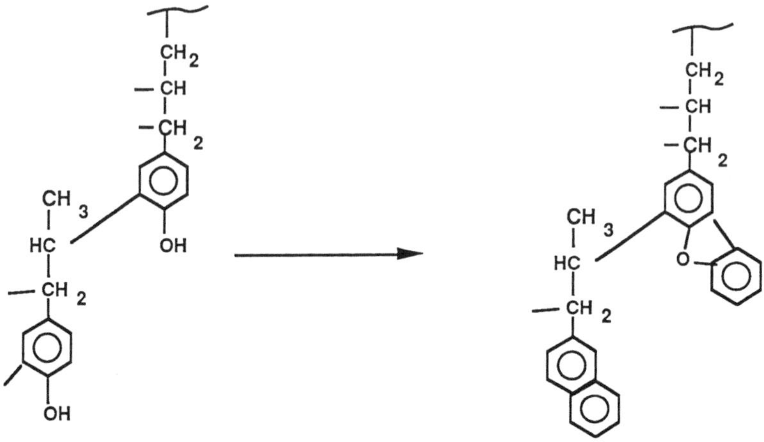

Figure 6. Proposed reactions responsible for the transformation of phenolic structures in subbituminous wood to dibenzofurans and naphthalenes.

sand bath for different reaction periods at various temperatures. Experiments were run for 10 days at 100° C, 30 minutes to 48 hours at 300° C, and 3 hours, 6 hours and 3 days at 350° C. After the elapsed heating times the reactor was removed and immediately quenched in water and then allowed to cool to room temperature. The gases, the liquids and the solids were collected and analyzed. Only trace amounts of gases were detected. An entire series of hydrous pyrolysis experiments were repeated at 300°C with more water added to the reactor than many of the previous experiments. This was done to ensure that water would remain in the liquid state throughout the experiment and would not be in the vapor state.

The bombs were opened and the liquid content pipetted off. The organic matter dissolved in the water was separated by extraction with diethylether, and the data will be reported at a later date. The residual lignite was extracted in a 50:50 mixture of benzene and methanol. The residue was then treated first with acetone and then with pentane to remove any benzene:methanol from the altered coal. The coal was then dried in a vacuum oven at 45° C for twenty four hours, and weighed.

Both the original lignite and the solid residues from the reactors were analyzed by flash pyrolysis and by solid state ^{13}C NMR. The flash pyrolysis technique used was that published by Hatcher et al. (*25*). Using a Chemical Data System Pyroprobe 1000, approximately one milligram of sample was loaded into a quartz capillary tube, and this tube was placed inside the coils of the pyroprobe. The probe and sample were then inserted into the injection port (temperature maintained at 280°C) of a Varian 2700 gas chromatograph and the sample pyrolyzed. The residue was first thermally desorbed at 300°C for thirty seconds. The samples were then pyrolyzed. Flash pyrolysis conditions were as follows: temperature, 610°C for 10 s with a heating rate 5°C/msec. The pyrolyzate was cryotrapped with liquid nitrogen prior to being chromatographed on a 25 m x 0.25 mm i.d. J & W DB-17 capillary column. The GC was temperature programmed from 30° to 280°C at 4°C/min. The effluent was swept into the source of a DuPont 490B mass spectrometer fitted with a Teknivent Vector/One data system for detection and compound identification. Compounds were identified by a combination of methods which included comparison of mass spectra to the NBS/Wiley library, to published mass spectra, and to authentic standards whenever possible.

Solid-state ^{13}C NMR spectra were obtained by the method of cross polarization and magic angle spinning (CPMAS) using the conditions previously given (*16*). The spectrometer was a Chemagnetics Inc. M-100 spectrometer operating at 25.2 MHz carbon frequency. Cycle times of 1sec and contact times of 1msec were chosen as the optimal conditions for quantitative spectroscopy.

Results

Elemental Analyses and Product Yields. The hydrous pyrolysis of the Patapsco lignitic wood induced pronounced alteration of the sample as depicted by the elemental compositions and product yields for some of the experiments shown in Tables I and II, respectively. Elemental compositions are only available for the original unaltered sample and three of the hydrous pyrolysis residues. The data indicate a general loss of hydrogen, relative increase in carbon and loss of oxygen as a function of increasing thermal severity that are all consistent in only a general way with trends during increased coalification. Product yields measured as the loss of sample weight during hydrous pyrolysis vary from a low of 8.4% to a high of 40%, with the products being mostly water soluble phenols and catechols (Wenzel and Hatcher, unpublished data).

Table I. Elemental data and vitrinite reflectance for the Patapsco log
and its residues obtained from hydrous pyrolysis.

sample	rank	R_0	%C	%H	%N	%O	H/C	O/C
Patapsco lignite	lignite	0.269	65.5	4.8	.34	28.4	0.873	0.326
residue 10 days 100° C	lignite*	0.296	69.2	4.4	.33	26.0	0.758	0.282
residue 0.5 hr 300° C	subbit.*	0.496	73.2	4.3	.22	22.3	0.700	0.229
residue 2 hr 300° C	HvC bit.*	na	76.9	3.4	.01	19.7	0.527	0.192
residue 3 hr 350° C	Mv. bit.*	1.37	na	na	na	na	na	na

*= rank obtained due to thermal alteration
na = data not available
R_0 = vitrinite reflectance, mean maximum (from Wenzel, unpublished data)

Table II. Hydrous pyrolysis yields at various temperatures.

time, hrs	temperature, °C	+conversion %
240	100	29.3
0.5	300	29.8(8.4)*
2	300	30.6(22.9)
3	300	(22.9)
5	300	(22.4)
8	300	(29.9)
3	350	39.7

*values in parenthesis are for samples to which more water
was added to ensure hydrous pyrolysis
+refers to the yield of non-residue

^{13}C NMR. The NMR spectra of Patapsco lignite and its hydrous pyrolysis residues
at different temperatures and heating times are shown in Figure 7. The unaltered
Patapsco lignitic wood has an NMR spectrum dominated by aromatic carbons with a
major peak at 145 ppm that is assigned to aryl-O carbons in catechol-like structures.
The aliphatic region (0-85 ppm) shows peaks assigned to alkyl-C carbons(0-50 ppm),
methoxyl carbons (56 ppm), and alkyl-O carbons (60-80 ppm). Two small broad
peaks at 170- 200 ppm are assigned to carboxyl and carbonyl carbons. The most
obvious change due to severity of hydrous pyrolysis is the diminution of the catechol
peak at 145 ppm. The most severely altered residue shown in Figure 7, 300°C for 48
hrs, shows a nearly complete loss of this peak; the peak of significance that emerges is
the one at 155 ppm, that of phenolic carbon.
    The other major change occurring with increasing hydrous pyrolysis severity
is the loss of aliphatic carbons relative to the aromatic carbons. Accordingly, the
carbon aromaticity, f_a, in Table III shows this trend as f_a increases from 0.68 to
0.88. The diminution of aliphatic carbons (Figure 7) involves loss of methoxyl, alkyl-
O, and alkyl-C carbons. The methoxyl carbons decrease rapidly but appear to persist
at 300°C. The alkyl-O carbons are lost at low thermal severity. It is clear that a
temperature of at least 350°C is required to reduce the methoxyl carbons to below
detection. The alkyl-C carbons show resonances distributed over a large chemical shift
range with an apparent shift in the intensity maximum from about 30 ppm to about 20
ppm. This is consistent with the conclusion that methyl carbons, which resonate at

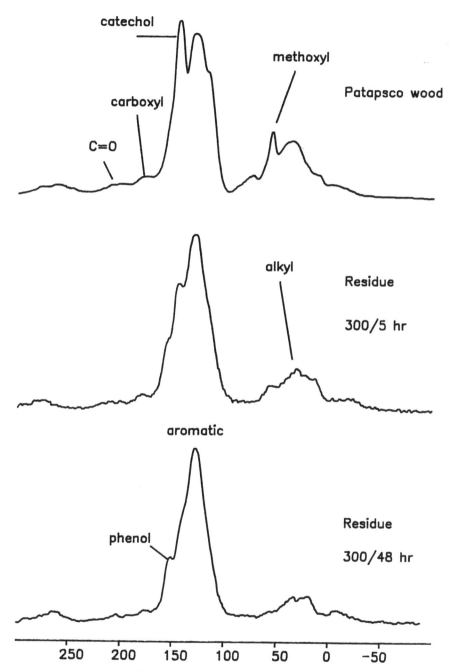

Figure 7. Solid-state ^{13}C NMR spectra of unaltered Patapsco lignitic wood and its hydrous pyrolysis residues heated at the various temperatures and times indicated.

lower chemical shifts, are dominant in the thermally stressed lignitic wood. Perhaps the residual aliphatic carbons are simply methyl carbons. The NMR data for samples subjected to hydrous pyrolysis at 300°C for intervals varying from 0.5 to 48 hrs and with additional water in the reactor are shown in Table III. The overall trends in this data are similar to those described above. Carbon aromaticity is observed to increase steadily over the course of 48 hrs of heating. Inverse trends are observed for aryl-O carbons (140-160 ppm), aryl-O/ aryl (aryl-O carbons normalized to total aromatic carbons), and aliphatic carbons (0-50 ppm). These trends are consistent with what was noted above-that hydrous pyrolysis induces loss of catechols and aliphatic structures. The precipitous loss of aryl-O carbons at the long heating time indicates additional loss of oxygenated functional groups, perhaps representing the condensation of phenols as is observed at higher coal rank (24).

Table III. NMR parameters for the Patapsco lignitic wood and hydrous pyrolysis residues measured by integration of the spectra.

Sample	f_a	Aryl-O/Aryl	#C=O	#COOH	$f_a^{a,H}$
Patapsco lignitic wood	0.68	0.40	2.6	3.2	0.41
residues/exp. 1:					
300°C/0.5hr	0.75	0.35	2.0	2.5	na
300°C/2hr	0.75	0.25	1.7	1.7	na
300°C/ 3hr	0.80	0.12	2.7	2.9	na
350°C/3hr	0.83	0.16	<0.5	2.4	0.27
350°C/72hr	0.88	0.12	<0.5	<0.5	na
°residues/ exp. 2:					
300°C/5hr	0.74	0.30	3.1	2.1	na
300°C/8hr	0.77	0.26	1.8	2.0	na
300°C/12hr	0.75	0.28	1.4	1.8	na
300°C/36hr	0.80	0.27	1.8	1.8	na
300°C/48hr	0.81	0.13	1.3	3.0	na

°in these experiments more water was added to reactor
# values are % of total carbon
na = not analyzed
f_a = carbon aromaticity
$f_a^{a,H}$ = fraction of aromatic carbons which are protonated

Dipolar dephasing of the unaltered Patapsco lignite (16) indicates that approximately 3.5 substituents are attached to the aromatic ring ($f_a^{a,H}$, or the fraction of protonated aromatic carbons, is 0.41). The value of $f_a^{a,H}$ for the residue stressed at 350°C for 3 hours is 0.27, significantly less than that of the original sample. This decrease in $f_a^{a,H}$ is indicative of increased aromatic substitution resulting from thermal stress.

PY/GC/MS

Flash pyrolysis GC/MS was preformed on the Patapsco lignite and the residues generated from the hydrous pyrolysis experiments in an effort to characterize the chemical changes that occured during alteration of the coalified wood. The changes in the lignite are best represented by comparing the pyrograms of the raw lignite (Figure 8) and of the series of altered residues from hydrous pyrolysis (Figure 9). The pyrograms are normalized to the highest peak. Major peaks are identified by number.

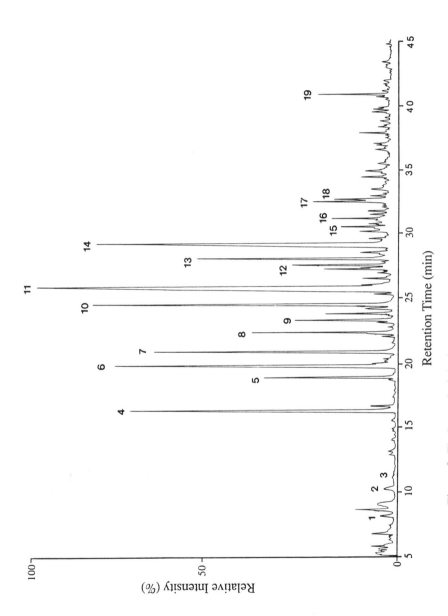

Figure 8. Flash pyrolysis/gc/ms trace of the unaltered Patapsco lignitic wood sample with identifications for major numbered peaks listed in Table IV.

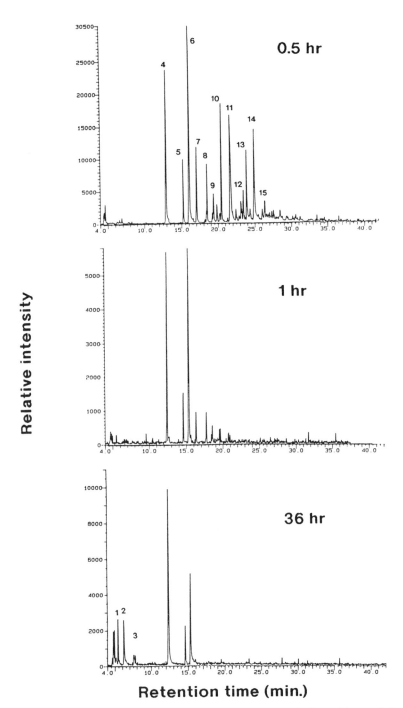

Figure 9. Flash pyrolysis/gc/ms trace of hydrous pyrolysis residues of the Patapsco lignitic wood heated for the various times indicated at 300°C. The numbered peaks are identified in Table IV.

See Table IV for the complete list of identified compounds for all the pyrograms. The pyrolyzate of the Patapsco coalified wood is dominated by peaks attributable to residual lignin structures as indicated by the presence of guaiacol, methylguaiacol, ethylguaiacol, and other methoxyphenols. The intense signals for catechol and alkylcatechols confirms the existence of a predominantly catechol-like structure for this sample. Phenol and alkylated phenols are present but in lesser amounts. Some of the phenols derive from thermal decomposition of catechol-like structures and lignin-like structures.

Pyrograms for the samples subjected to hydrous pyrolysis are distinctly different from those of the unaltered lignitic wood. With increasing hydrous pyrolysis severity, a significant diminution of catechols and lignin phenols (guaiacols) is observed. The pyrograms become dominated by phenol and its alkylated homologs, in much the same way which occurs naturally during coalification of wood. However, the major difference compared to naturally coalified wood is the general lack of significant quantities of the alkylated phenolic homologs. The pyrolysis of the residues yields mostly phenol and cresol isomers. Other alkylated phenols, the C_2 and C_3 phenols, are not as abundant relative to phenol and cresols as are the C_2 and C_3 phenols in the pyrolysis of the original Patapsco lignite or in coalified wood from a natural coalification series *(24)*. The relative depletion of C_2 and C_3 alkylphenols in the artificial series is probably related to the fact that alkyl substitutents are not abundant in the residue and the alkyl substituents are probably mostly methyl substituents as was deduced from NMR data.

Table IV. Peak assignments for the pyrograms in Figures 8 and 9.

Peak #	assignment
1	toluene
2	xylene
3	xylene
4	phenol
5	*ortho*-cresol
6	*meta and para*-cresol
7	guaiacol
8	2,4-dimethylphenol
9	4-ethylphenol
10	4-methylguaiacol
11	catechol
12	C_2-guaiacol
13	3-methylcatechol
14	4-methylcatechol
15	C_4-guaiacol
16	C_2-catechol
17	ethylcatechol
18	C_2-naphthalene
19	C_5-naphthalene

Figure 10 shows the trends in the compound yields from the flash pyrolysis products of hydrous pyrolysis residues heated at 300°C for varying periods of time. Compounds are grouped according to five major groups. The groups of compounds are benzene, phenol, guaiacol, catechol and naphthalene and their alkylated derivatives. The loss of catechol-like structures is documented with a significant loss of catechols in the pyrolyzate of the altered residues. With the loss of catechols, phenols and alkyl phenols become the dominant products in the pyrolyzate, as noted above.

Relative Yields of Flash Pyrolysis Products from Hydrous Pyrolysis Residues

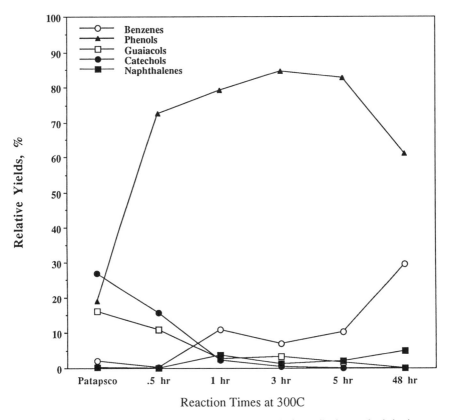

Figure 10. Relative yields of individual compounds from flash pyrolysis/gc/ms of hydrous pyrolysis residues obtained at 300°C plotted as a function of heating time (hrs). The compounds are grouped into compound classes and include all the various alkylated isomers of each class.

Discussion

Hydrous pyrolysis of the Patapsco lignite was conceived as a means to induce artificial coalification. Accordingly, our goal was to examine the residues before and after thermal treatment and to infer coalification pathways from this treatment. It is important to note that there appeared to be a loss of mass during thermolysis. This mass loss probably reflects low molecular weight products formed, water, and partial solubilization of lignitic structures originally of low molecular weight. The GC/MS analysis of the water soluble materials (Hatcher and Wenzel, unpublished results) indicates the presence of lignin phenols and catechols, all of which are components of the lignite prior to thermolysis. Cleavage of a few bonds on the macromolecule could conceivably produce a catechol or methoxyphenol which would be soluble in water. Another water soluble product identified was acetic acid. This product is most likely derived from hydrolysis of the side-chain carbons associated with the original lignin and coalified lignin in the sample. The content of evolved gases was not quantified because the amount of gas evolved was negligible.

The van Krevelen diagram (Figure 11) best represents the elemental changes that occur in the naturally coalified wood samples versus the elemental changes that occur in the artificially altered coalified wood samples. With increasing maturation, the H/C ratio for vitrinite remains relatively flat while the O/C value decreases steadily until HvA bituminous at 0.02 O/C, (approx. 88 % carbon). In the artificial series there is an immediate loss of both hydrogen and oxygen. The hydrogen content for the Patapsco coalified wood is initially 4.8 percent and this value decreases to a low in the most altered residue of 3.4 percent. It is quite evident that the artificial thermal pathway does not follow the natural coalification pathway of vitrinites on the van Krevelen diagram (*26*). Not enough oxygen is lost with respect to hydrogen. The trend appears to follow one of dehydration which involves loss of 2 hydrogens for each oxygen.

The NMR data of the thermolysis residues indicates clearly the evolutionary path during thermolysis. Comparing the NMR data for the residues (Figure 7, Table III) with the original lignite and with gymnospermous wood coalified to higher rank (*16*) there appear to be several changes which describe the average chemical alteration. First, the most obvious change is the loss of aryl-O carbons having a chemical shift at 145 ppm. From previous studies (*16*) this peak has been assigned to aryl-O carbons in catechol-like structures in the coalified wood, based on chemical shift assignments and pyrolysis data. These are thought to be originally derived from demethylation of lignin during coalification. The NMR spectrum of the hydrous pyrolysis residue has clearly lost most of the intensity at 145 ppm but now shows a peak at 155 ppm which is related to aryl-O carbons in monohydric phenols. These are the primary constituents of subbituminous and high volatile bituminous coalified wood as depicted in the NMR spectra for such woods (*24*). It is obvious that the catechol-like structures of the lignitic wood have been tranformed to phenol-like structures somewhat similar to those in higher rank coal. Thus, the hydrous pyrolysis has reproduced, to some degree, the coalification reactions acting on aromatic centers.

The second most apparent change to occur during hydrous pyrolysis is the loss of aliphatic structures. The methoxyl groups at 56 ppm are lost from the lignitic wood as are the other alkyl-O carbons at 74 ppm, consistent with demethylation reactions and dehydroxylation of the three carbon side chains of lignin which occur naturally during coalification of woods. However, the loss of alkyl-C carbons (those aliphatic carbons not substituted by oxygen) is the most significant change in the aliphatic region in the hydrous pyrolysis residue. Loss of substantial amounts of aliphatic carbon is not observed during coalification of wood from lignite to high volatile bituminous coal. In fact, in most coalified wood samples, loss of alkyl-C carbon occurs only at higher ranks, above that of medium volatile bituminous coal. The lack

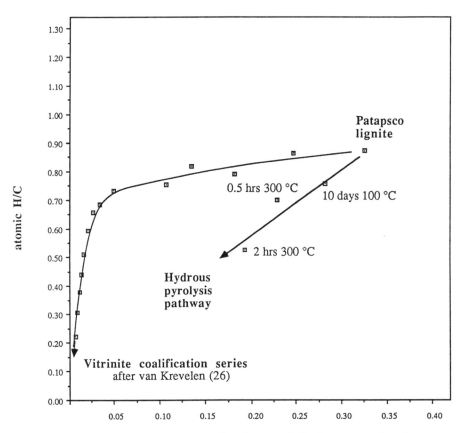

Figure 11. van Krevelen diagram showing trends in the nature of hydrous pyrolysis residues and a natural coalification series.

of retention of aliphatic carbon during hydrous pyrolysis is an indication that this treatment probably does not reproduce well the low-rank coalification reactions associated with aliphatic structures. It is important to note that the alkyl-C carbons in the hydrous pyrolysis residues become dominated by methyl carbons with increasing thermal stress.

The pyrolysis data provide confirmation for the average changes in structure observed by NMR. The loss of catechol-like structures is documented with the significant diminution of catechols in the pyrolyzate of the hydrous pyrolysis residue compared to that of the original lignitic wood. This loss of catechols is the singular most significant change in pyrolysis products. The pyrolyzate of the residue mimics somewhat the pyrolysis data for subbituminous coalified wood (*24*), being dominated by phenol and alkylphenols. Another difference between hydrous pyrolysis residues and the original lignitic wood is in the abundance of methoxyphenols derived from lignin-like structures. The hydrous pyrolysis has apparently reduced the relative yields significantly, consistent with the NMR data showing loss of methoxyl carbons. Some of the methoxyphenols may have been transformed to water soluble phenols and washed out of the residue in the aqueous phase; others may have undergone demethylation reactions, converted to catechols and then transformed to phenols.

The pyrolysis of the residues yield mostly phenol and the cresol isomers; other alkylated phenols, the C_2 and C_3 phenols, are not as abundant relative to phenol and cresols as the C_2 and C_3 phenols are in the pyrolysis of original lignite or coalified woods of higher rank (*24*). This is probably related to the fact that alkyl substituents are not as abundant in the residue and the alkyl substituents are probably mostly methyl substituents as was deduced from the NMR data. Thus, the lack of significant relative amounts of C_2 and C_3 phenols in the thermolysis residue's pyrolyzate further supports the conclusion that thermolysis does not reproduce coalification reactions with regard to the aliphatic structures in the residue. The relatively high temperatures used in this study may force proportionally more thermolytic pathways over ionic pathways. The potential for such a situation has been recognized by Siskin et al. (*12*).

By noting the changes in the NMR and pyrolysis data when lignitic wood is subjected to hydrous pyrolysis and by comparing the residue thus formed with coalified wood of subbituminous rank, a generalized schematic of the transformation pathway may be deduced. It is clear that the catechol-like structures are being transformed to phenolic structures. From previous studies on the lignitic wood, catechol-like structures are thought to be the dominant structural element, similar to the one depicted in structure **1**. Thermolysis can easily induce dehydration or condensation (*10*) to form an ether link between two catechols as shown in structure **2**. Hydrolysis of the ether can reverse the reaction to regenerate the catechols, but thermolysis can also induce homolytic cleavage of the ether to generate a phenoxy radical (**3**) and a catechol radical (**4**). If we assume that sufficient hydrogen transfer occurs under the high temperature regime to cap the radicals, then the products **3** and **4** would be tranformed to a phenol (**5**) and a catechol (**1**). It is also possible that radicals are capped by methyl or alkyl radicals formed from other thermolytic reactions elsewhere in the coal structure (e.g., the side chain sites). In such cases methoxy or alkoxyphenols (**6**) could be formed. Such reactions could explain why methoxyphenols persist in pyrolyzates of the residues during hydrous pyrolysis. It is also possible that the alkyl radicals could attack the aromatic ring sites and induce alkylation (**7**). This would lead to increased substitution of the rings, consistent with the dipolar dephasing NMR data which show an increase in the average number of aromatic substituents with increased hydrous pyrolysis.

Recently, Siskin et al.(*12*) demonstrated that hydrous pyrolysis at 343°C will induce cleavage of 4-phenoxyphenol (**8**) to form phenol instead of hydroquinone (1,4-dihydroxybenzene); the detailed pathway being presumably unknown. This compound has some similarity to structure **2** found in the lignitic wood.

Thus we can expect the above mechanism proposed for catechol to be analogous to that shown for 4-phenoxyphenol. In fact, we might expect the reaction to proceed at a much faster rate due to the fact that additional phenol substituents at *orto* or *para* positions to the ether linkage can activate the reaction.

Acknowledgments

We thank The Pittsburgh Energy Technology Center and M. Nowak for partial financial support through Contract DE-AC22-91PC91042.

Literature Cited

1. Dyrkacz, G. R.; Horwitz, E. P. *Fuel* **1982**, *61*, p. 3.
2. Dyrkacz, G. R.; Bloomquist, C. A. A.; and Ruscio, L. *Fuel* **1984**, *63*, p. 1367.
3. Greenwood, P. F.; Zhang, E.; Vastola, F. J.; Hatcher, P. G. *Anal. Chem.* **1993**, *65*, p. 1937.
4. Stout, S. A.; Hall, K. J. *J. Anal. Appl. Pyrol.* **1990**, *21*, p. 195.
5. Hatcher, P. G.; Lerch, H.E.; Verheyen, T. V.; *Int. Jour. Coal. Geol.* **1989**, *13*, p. 65.
6. Hatcher, P. G.; Wilson, M.A.; Vassolo, A. M.; *Int. Jour. Coal. Geol.* **1989**, *13*, p. 99.
7. Hatcher, P. G. *Org.Geochem.* **1990**, *16*, p. 959.
8. Hatcher, P. G.; Breger, I. A.; Szeverenyi, N.; Maciel, G. E. *Org. Geochem.* **1982**, *4*, p. 9
9. Hedges, J.J.; Cowie, G.L.; Ertel, J.R.; Barbour, R.J.; Hatcher, P.G. *Geochim.Cosmochim. Acta* **1985**, *49*, p. 701.
10. Siskin, M.; Katritzky, A.K. *Science* **1991**, *254*, p.231.
11. Katrikzky, A.K.; Murugan, R.; Balasubramanian, M.; Siskin,M.; *Energy Fuels* **1990**, *4*, p. 543.
12. Siskin, M.; Brons, G.; Vaughn, S.N.; Katritzky, A.K.; Balasubramanian, M.;*Energy Fuels* **1990**, *4*, p. 488.
13. Bates, A. L.; Hatcher, P. G.*Org. Geochem.***1989**, *14*, p. 609.
14. Hatcher, P. G.; Lerch, H. E.; Bates, A. L.; Verheyen, T. V.; *Org.Geochem.* **1989**,*14*, p. 145.
15. Stout, S. A.; Boon, J. J.; Spackman, W. *Geochim. Cosmochim. Acta* **1988**, *52*, p. 405.
16. Hatcher, P.G. *Energy Fuels* **1988**, *2*, 4p. 8.
17. Ohta, K.; Venkatesan, M. I. *Energy Fuels* **1992**, *6*, p. 271.
18. Botto, R. E. *Energy Fuels* **1987**, *1*, p. 228.
19. Adler E. *Wood Sci. Technol.* **1977**, *11*, p. 169.
20. Nimz H. *Angew. Chem Int. Ed. Engl.* **1974**, *13*, p. 313.
21. Faulon, J.-L.; Hatcher, P. G., *Energy Fuels* (in press).
22. Hatcher, P. G.; Wenzel, K. A. ; Faulon, J.-L., Preprints, American Chemical Society Fuel Division **1993**, *38*, p. 1270.
23. Stach, E.; Taylor, G. H.; Mackowski, M-Th.; Chandra, D.; Teichmüller, M. ; Teichmüller, R. *Stach's Texbook of Coal Petrology*; Gebrüder Borntrüger, Berlin, 1975; 428pp.
24. Hatcher, P. G.; Faulon, J.-L.; Wenzel, K. A. ; Cody, G. D. *Energy Fuels* **1992**, *6*, p. 813.
25. Hatcher, P. G.; Lerch, H. E.; Kotra, P.K.; Verheyen, T. V.; *Fuel* **1988**, *67*, p. 106.
26. Wenzel, K. A.; Hatcher, P. G.; Cody, G. D.; Song, C., Preprints, American Chemical Society Fuel Division **1992**, *38*, p. 571.
27. van Krevelen, D. W. *Fuel* **1950**, *29*, p. 269.

RECEIVED April 15, 1994

Chapter 9

Flash Pyrolysis–Gas Chromatography–Mass Spectrometry of Lower Kittanning Vitrinites

Changes in Distributions of Polyaromatic Hydrocarbons as a Function of Coal Rank

Michael A. Kruge and David F. Bensley

Department of Geology, Southern Illinois University, Carbondale, IL 62901

Chemical analyses restricted to a single coal maceral permit a focus on rank effects without concern for variations in organic matter type. Vitrinite concentrates of high purity were isolated from coal samples of the Lower Kittanning seam by multi-step density gradient centrifugation, with reflectances ranging from 0.66 to 1.39% R_{max}. In addition to the previously recognized losses of phenolic compounds, the vitrinite pyrolyzates exhibit marked increases in relative concentrations of tri- and tetraaromatic hydrocarbons (especially benzo[a]fluorene, methyl-phenanthrenes, methylfluorenes and methylchrysenes) above 0.9% R_{max}, i.e., beyond the "second coalification jump" of Teichmüller (1). Thus, petrographically-recognizable physical transformations are shown to correspond to a major chemical restructuring of the vitrinite. Increases in rank also correspond to systematic variations in the distributions of the isomers of various polyaromatic compounds in the flash pyrolyzates, which can be exploited in the creation of indicators covering a wide range of thermal maturation.

Most attempts to generalize the molecular structure of bituminous coal have relied upon the concepts of clusters, that is, sub-units with a small number of fused aromatic ring systems connected by short aliphatic chains and heteroatom bridges (e.g., 2). The sub-units are conceived as "average" or "representative" of the larger coal structure and are convenient, as they can be represented graphically in two dimensions on a printed page. Such models are the results of synthesis of data from multiple sources, including elemental analysis, chromatographic and mass spectrometric evaluation of pyrolysis and chemolysis products, and spectroscopic studies, in particular solid state nuclear magnetic resonance (NMR) and infra-red (IR). These concepts can be extended using computerized three-dimensional modeling, which reveals likely interactions between the hypothetical sub-units, permitting the determination of the most energetically-favorable conformations (3). Through such efforts, there has been a recognition of the role of non-covalent bonding in the formation and stabilization of the 3-D structure, with hydrogen bonding important when functional groups containing oxygen are present and van der Waals forces serving to bind stacked adjacent aromatic ring structures (*3, 4*). However, in spite of their sophistication, such models suffer from an obvious deficiency in that they attempt to present a structure for whole coal, which is in fact a variable mixture of solid macerals of diverse biological origin, hence of differing molecular structures from the outset (*5*), plus a complex mobile phase (*6, 7*). Flash pyrolysis studies of a variety of macerals isolated from the same coal or kerogen sample have shown that there are significant chemical differences among such macerals (*8-13*). In addition, the profound effects of increasing coal rank on chemical structure are not often included in molecular modeling attempts. Each maceral may be expected to react differently to thermal alteration, just as different organic matter types follow distinct pathways on van Krevelen diagrams (*14*). Recent studies have focused on the transition from lignin to fossil wood (vitrinite) of high volatile bituminous rank, thus following the evolution of a single maceral type through the early stages of coalification (*5, 15, 16*), using the lignin structure as the starting point for the development of a model of vitrinite.

The application of the medical technique of density gradient centrifugation (DGC) to sedimentary organic matter (*17*) permits the preparation of maceral concentrates in quantities suitable for chemical analysis. DGC is particularly effective in isolating vitrinite from humic coal (Bensley, D. F.; Crelling, J. C., *Fuel*, in press). The Upper Carboniferous Lower Kittanning seam of Pennsylvania and Ohio (USA) is a single coal bed, which if followed laterally, ranges in rank from sub-bituminous to low volatile bituminous (*18*). It thus provides an ideal rank series to monitor the effects of thermal alteration over the entire oil generation window. By examining pyrolyzates of high purity vitrinite concentrates from Kittanning samples, it should be possible to isolate the effects of increasing coal rank on the molecular structure of a single maceral type. This work is thus, in part, an extension of previous studies (*5*) into a higher rank range.

Experimental Methods

Vitrinite concentrates were prepared using four coals of different rank from the Lower Kittanning seam, obtained from the Pennsylvania State University sample bank (Table I). The coals were processed by mechanical crushing to <75 μm, and (after pre-treatment with liquid nitrogen to induce fracturing) by micronization using a fluid energy mill employing nitrogen carrier gas and a 3 mL/min. charge rate. Overall density properties of the samples were determined by density gradient centrifugation (DGC) of 2 g of micronized coal over a density range of 1.0 to 1.6 g/mL, in an aqueous solution of CsCl, the concentration of which varied systematically (*17*). Once the density ranges of the vitrinite group macerals were determined (Table I), fresh aliquots of the micronized samples were successively centrifuged in CsCl solutions of constant density corresponding to the lower and upper limits of the vitrinite range. These pre-concentrated vitrinites were subjected to a final centrifugation step over the appropriately narrow density range and individual vitrinite fractions were produced. Extreme care was taken to avoid any chance of multi-maceral or non-vitrinite maceral contamination. Semifusinite contamination was evaluated by monitoring reflectance distributions. Blue-light analysis showed no trace of fluorescence in the vitrinite fractions, indicating that there was no liptinite contamination. The purity of the vitrinite fractions was petrographically determined to be >99%. Reflectance data was obtained using a Leitz MPVIII compact microscope modified for rotational polarization reflectance (*19, 20*). It should be noted that for each sample, a wide range of reflectances was apparent, after each vitrinite fraction was examined individually. This phenomenon is discussed in detail elsewhere (Bensley, D. F.; Crelling, J. C., *Fuel*, in press).

Table I. Samples Employed and Their Reflectance and Density Values

Sample (PSU)[a]	Sample (SIU)[b]	Bulk R_{max} (%)[c]	Vitrinite Density Range (g/mL)	Density Fraction Used (g/mL)	Fraction R_{max} (%)
PSOC-1289	SIU-1904	0.67	1.27-1.33	1.30	0.66
PSOC-1142	SIU-1889	0.81	1.25-1.31	1.30	0.82
PSOC-1145	SIU-1891	1.14	1.25-1.32	1.30	1.02
PSOC-1325	SIU-1907	1.59	1.24-1.29	1.27	1.39

[a]The Pennsylvania State University coal sample bank number.
[b]Coal sample number assigned at Southern Illinois University.
[c]Average vitrinite reflectance, measured at SIU on the whole coal sample.

For each of the four samples, an aliquot of the fraction corresponding to the approximate mode of the vitrinite DGC peak (Table I), pre-extracted with CH_2Cl_2, was subjected to analytical pyrolysis-gas chromatography/mass spectrometry (py-GC/MS).   Pyrolysis was at 610° C for 20 sec., using a CDS 120 Pyroprobe coupled to an HP 5890A GC and an HP 5970B Mass Selective Detector. The pyrolysis temperature was monitored by a thermocouple situated inside the sample holder. The GC was equipped with a 25 m OV-1 column (0.2 mm i.d., 0.33 μm film thickness), initially held at 0° C for 5 min., then raised to 300° at 5°/min., then held for 15 min. The mass spectrometer was in full scan mode with an ionizing voltage of 70 eV.

Quantitations of the aromatic compounds of interest were performed on the Hewlett Packard data system, using the mass chromatograms of their molecular ions. All ratios used as indicators of the degree of thermal alteration are arranged such that the thermally stable components are in the numerator and the sum of the stable and labile are in the denominator. Thus all ratios increase with rank and are constrained to vary between 0 and 1. The exception is the Methylphenanthrene Index (*21*), which is used in its published form. Note that the plural forms of aromatic compound names are used to collectively refer to the principal compound and all detected pseudohomologues. For example, the group term "naphthalenes" is used to mean naphthalene plus the C_1 to C_3-alkylnaphthalenes. Compounds are also collectively designated herein by upper case Greek letters followed by an Arabic numeral, signifying the compound type(s) and the extent of methyl substitution (Figure 1). For example, "N$_3$" refers to C_3-alkylnaphthalenes and "Ψ_2" refers to compounds with a molecular ion of m/z 206, i.e., C_2-alkylphenanthrenes and C_2-alkylanthracenes.

Results and Discussion

Distributions of Aromatic Compounds by Class.   The four vitrinite pyrolyzates all show a predominance of aromatic hydrocarbons and/or phenolic compounds. There are only minor amounts of aliphatic hydrocarbons, due to the absence of liptinites and extractable organic material. Phenol and C_1 to C_3-alkylphenols are the most abundant compounds for the 0.66% R_{max} vitrinite (Figures

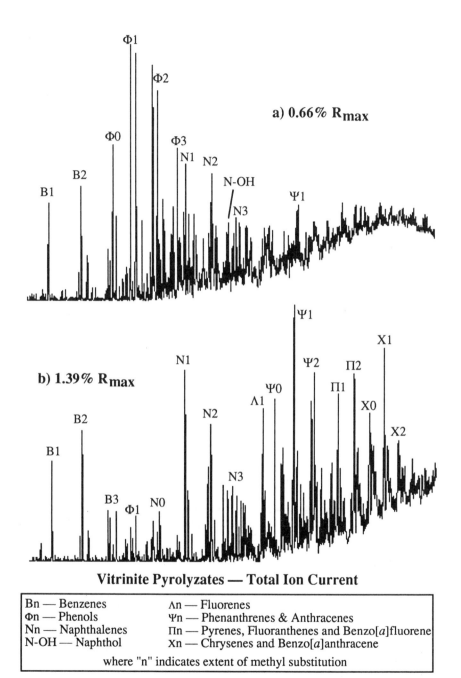

Figure 1 — Py-GC/MS total ion current chromatograms of solvent-extracted vitrinite concentrates. a) Low rank example. b) High rank example.

1a and 2). In both relative and absolute terms, concentrations of the phenols diminish with increasing rank and thus, at 1.39% R_{max} they are only of minor importance (Figures 1b and 2), a phenomenon previously noted (22, 23). Naphthol, another aromatic alcohol, is also detectable as a significant compound in the 0.66% R_{max} pyrolyzate (Figure 1a), but not in the higher rank samples. The presence of hydroxyl functions in low rank vitrinites and their subsequent loss due to thermal alteration fits the well-documented decrease in the O/C atomic ratio for Type III organic matter as rank increases (14). It is the continuation of the trend of the loss of oxygen functionalities observed in the transition from lignin to sub-bituminous coal (5,15,16).

Mono- and Diaromatic Hydrocarbons. Monoaromatic hydrocarbons (C_1 to C_3-alkylbenzenes in particular) are important compounds for all vitrinite samples, regardless of rank (Peaks B_1-B_3 in Figure 1). Concentrations of monoaromatic hydrocarbons relative to other aromatics exhibit a minor decrease with rank (Figure 2). Benzene is detectable as a minor peak in all four pyrolyzates, but its low concentrations are in part an artifact of the analytical conditions employed. Therefore, it should be noted that Figure 2 presents the quantitation results for unsubstituted plus monomethyl aromatic compounds, except in the case of the benzene series, for which methyl- plus dimethylbenzenes are used. Diaromatic hydrocarbons (naphthalene and C_1 to C_3-alkylnaphthalenes) are important components in all the pyrolyzates, especially the methyl- and dimethylnaphthalenes (Peaks N_1 and N_2 in Figure 1). Slightly higher relative concentrations of naphthalene plus methylnaphthalenes are noted for the vitrinites in the mid-rank range (0.82 and 1.02% R_{max}, Figure 2).

Triaromatic hydrocarbons. Triaromatic hydrocarbons are minor, although readily detectable, constituents of the low rank vitrinite pyrolyzate, with methylphenanthrene and methylanthracene the most notable (Ψ_1 in Figure 1a). In contrast, desmethyl-, methyl- and dimethylphenanthrenes are the most important peaks on the total ion current trace of the high rank pyrolyzate (Figure 1b). Relative and absolute concentrations of phenanthrene plus methylphenanthrenes increase sharply with rank, especially between the 0.82 and 1.02% R_{max} samples (Figure 2). The concentrations of fluorenes, in particular methylfluorene, are also greatly enhanced at high rank (Peak Λ_1 in Figure 1b).

Tetraaromatic Hydrocarbons. Fluoranthene, pyrene, benzo[a]fluorene, chrysene, benzo[a]anthracene and methylated derivatives are detectable, albeit faintly, in the lower rank pyrolyzates. The higher rank samples present a radically different picture, in which the tetraaromatics are among the major components, chrysenes and benzo[a]fluorene in particular (Figure 1b). The relative and absolute concentrations of "Group II" compounds, including pyrene, fluoranthene, methylpyrenes and benzo[a]fluorene (i.e., those having molecular ions of m/z 202 and 216) increase sharply between 0.82 and 1.02% R_{max}, much like the phenanthrenes (Figure 2). The "Group X" compounds (those with molecular ions at m/z 228 and 242: chrysene, benzo[a]anthracene and their methylated derivatives) behave in an analogous fashion (Figure 2). The "second coalification jump" as defined by Teichmüller (1), occurring at ≈0.90% R_{max}, marks petrographically-recognizable physical changes in coal and the end of the *peak* stage of hydrocarbon generation. This boundary is shown here to correspond to profound changes in chemical structure of vitrinite as well, as seen in

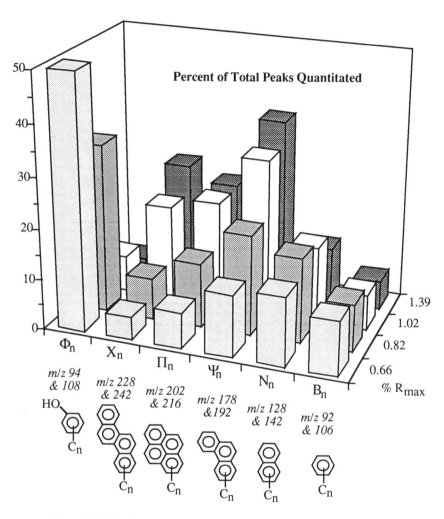

Figure 2. Relative quantitation results for key aromatic and phenolic compounds in the flash pyrolyzates of four vitrinite concentrates. Compound groups are designated by upper case Greek letters as in Figure 1. Molecular ions used in quantitation are given. Examples of molecular structures are also shown, but it should be noted that "Group X" also includes benzo[a]anthracene, "Group Π" also includes fluoranthene and benzo[a]fluorene, and "Group Ψ" also includes anthracene. Subscript "n" indicates the extent of methyl substitution (n = 0 and 1 for all compounds except the benzene series, for which n = 1 and 2). See text for further explanation.

the sharp "chemical divide" visible between the 0.82 and 1.02% R_{max} samples in Figure 2. The py-GC/MS results provide details of the evolution of the molecular structure of vitrinite with rank, going beyond the simple "increasing degree of condensation" previously inferred from solid state NMR data (*e.g.*, *7, 24*). The tri- and tetraaromatic hydrocarbons clearly dominate the pyrolyzates of the higher rank samples. These data suggest that the predominance of the larger aromatic structures is the result of progressive condensation (accelerating beyond 0.90% R_{max}) of smaller units, mostly phenolic structures as they undergo reduction.

Distributions of Polyaromatic Hydrocarbon Isomers. Profound rank-induced changes have been demonstrated in distribution of the aromatic compounds by number of rings, extent of methyl and hydroxyl substitution, and ring configuration. It is also of interest to examine the samples for variations in the distributions of key isomers of polyaromatic hydrocarbons (*25, 26*). The isomer clusters of interest can be observed on mass chromatograms of the appropriate molecular ion. Peaks can be quantitated and ratios calculated which may conveniently mark the progress of increasing thermal alteration. Such an approach is most frequently used on aromatic fractions of petroleums and source rock extracts. It is also advantageous to be able to apply such methods to flash pyrolyzates (*27*).

Trimethylnaphthalenes. The effect of an increase in rank from 0.66 to 1.39% R_{max} on the distributions of trimethylnaphthalene isomers can be observed on m/z 170 mass chromatograms (Figure 3). The 1,3,7-, 1,3,6- and 2,3,6-trimethyl isomers (Peaks 1, 2 and 4) are clearly favored at high rank. These isomers have been recognized for their superior thermal stability in natural and artificially-matured samples, and also in light of theoretical considerations (*28-30*). Thus, their usefulness as indicators of the extent of thermal alteration may be expanded to include flash pyrolyzates. If the summed quantitation results of the three peaks are placed in the numerator, and those of all five trimethylnaphthalene peaks are in the denominator, the resulting "TMN" ratio increases slowly from 0.66 to 0.82% R_{max}, then more rapidly thereafter (Figure 4).

Phenanthrenes and Anthracenes. Distributions of phenanthrene, anthracene and their methylated variants also change significantly with rank, as seen on m/z 178 and 192 mass chromatograms (Figure 3). By 1.39% R_{max}, relative amounts of anthracenes and 9- and 1-methylphenanthrenes (Peaks 7, 10, 11 and 12) have decreased sharply. Methylphenanthrene Index or "MPI" (*21*) values show little change at lower ranks but increase sharply between 0.82 and 1.02% R_{max}, as the above-mentioned chemical divide is crossed (Figure 4). Equivalent vitrinite reflectance values (R_c, *25*) computed from the MPI of these experiments do not correspond well with the measured reflectances, especially at the lower ranks. This may be due to the coelution of a methylanthracene with 1-methylphenanthrene in Peak 12, which would have the effect of suppressing the ratio.

The anthracenes in the pyrolyzates are present in relative concentrations well in excess of what would be expected for petroleums and rock extracts (even coal extracts) of comparable rank. Since their concentrations are clearly rank dependent (Figure 3), they may also be exploited in the creation of thermal alteration indicators, applicable to vitrinite pyrolyzates. A simple ratio ("PA") of phenanthrene to the sum of anthracene and phenanthrene (Peaks 6/(6+7)) increases linearly up to 1.02% R_{max}, beyond which

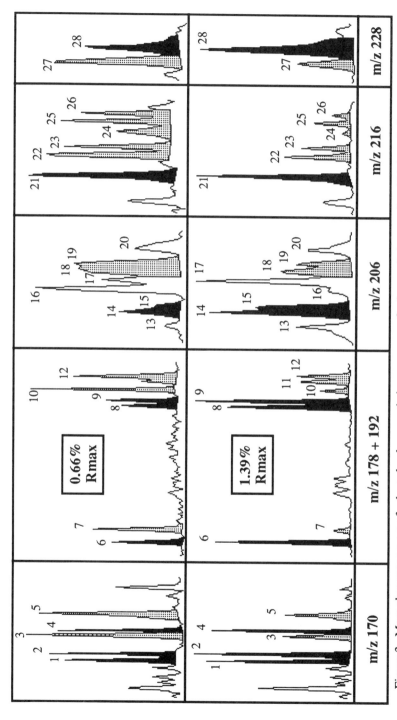

Figure 3. Mass chromatograms of selected polyaromatic isomer groups for low rank (top row) and high rank (bottom row) vitrinite concentrates. Compounds are identified in Table II. Peaks used in the thermal alteration indicator ratios of Figure 4 are shown in black if they are labile. See text for further explanation.

Table II — Identification of Peaks Used in Figure 3.

Peak	m/z	Group[a]	Compound(s)
1	170	N_3	1,3,7-trimethylnaphthalene
2	170	N_3	1,3,6-trimethylnaphthalene
3	170	N_3	1,4,6- & 1,3,5-trimethylnaphthalenes
4	170	N_3	2,3,6-trimethylnaphthalene
5	170	N_3	1,2,7- & 1,6,7- & 1,2,6-trimethylnaphthalenes
6	178	Ψ_0	phenanthrene
7	178	Ψ_0	anthracene
8	192	Ψ_1	3-methylphenanthrene
9	192	Ψ_1	2-methylphenanthrene
10	192	Ψ_1	methylanthracene
11	192	Ψ_1	9-methylphenanthrene
12	192	Ψ_1	1-methylphenanthrene & methylanthracene
13	206	Ψ_2	3,6-dimethylphenanthrene
14	206	Ψ_2	2,6-dimethylphenanthrene
15	206	Ψ_2	2,7-dimethylphenanthrene
16	206	Ψ_2	dimethylanthracene
17	206	Ψ_2	1,3- & 2,10- & 3,9- & 3,10-dimethylphenanthrenes
18	206	Ψ_2	1,6- & 2,9-dimethylphenanthrenes
19	206	Ψ_2	1,7-dimethylphenanthrene
20	206	Ψ_2	2,3- & 1,9- & 4,9-dimethylphenanthrenes
21	216	Π_1	benzo[a]fluorene
22	216	Π_1	methylfluoranthene or benzofluorene (?)
23	216	Π_1	2-methylpyrene
24	216	Π_1	methylfluoranthene or benzofluorene (?)
25	216	Π_1	4-methylpyrene
26	216	Π_1	1-methylpyrene
27	228	X_0	benzo[a]anthracene
28	228	X_0	chrysene

[a]For comparison with total ion chromatograms in Figure 1.

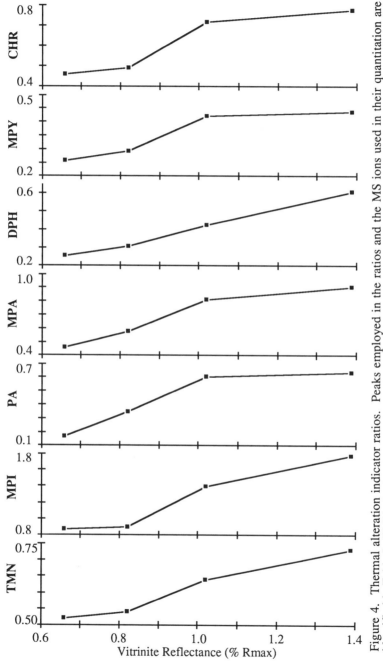

Figure 4. Thermal alteration indicator ratios. Peaks employed in the ratios and the MS ions used in their quantitation are identified in Table I. *TMN*: Peaks (1+2+4) / (1+2+3+4+5). *MPI*: (1.5 (Peaks 8+9) / (Peaks 6+11+12). *PA*: Peaks 6 / (6+7). *MPA*: Peaks (8+9) / (8+9+10). *DPH*: Peaks (14+15) / (14+15+18+19). *MPY*: Peaks 21 / (21+22+23+24+25+26). *CHR*: Peaks 28 / (27+28). See text for further explanation.

it stabilizes (Figure 4). This discontinuous behavior marks the crossing of the chemical divide. The 2- and 3-methylphenanthrenes and methylanthracene (Peaks 8, 9 and 10) are related in a similar fashion using the ratio "MPA" (Figure 4), which also exhibits a decrease in slope above 1.02% R_{max}.

The dimethylphenanthrenes (m/z 206, Figure 3) exhibit marked augmentation of the 2,6- and 2,7-dimethyl isomers (Peaks 14 and 15) as the vitrinite reflectance increases from 0.66 to 1.39% R_{max}. The thermal stability of these peaks has been previously noted, both by empirical observation and on a theoretical basis (25, 26, 31, 32). There is a coincident decrease observed for the less stable 1,6-, 2,9- and 1,7-dimethyl isomers (Peaks 18 and 19). These phenomena can be exploited in the creation of the "DPH" ratio (Figure 4), which increases smoothly throughout the entire rank range. The dramatic loss of dimethylanthracene (Peak 16, Figure 3) with rank could also be used as the basis for a ratio which would work for vitrinite pyrolyzates.

Tetraaromatic Hydrocarbons. Among "Π_1" compounds, benzo[a]fluorene (Peak 21, Figure 3) demonstrates a superior thermal stability over the methylpyrene isomers (Peaks 23, 25 and 26) seen on the m/z 216 mass chromatograms. Unidentified Peaks 22 and 24, which are probably methylfluoranthenes or other configurations of benzofluorene, also show relative decreases with rank. The resulting "MPY" ratio (Figure 4) has a sharp increase in slope between 0.82 and 1.02% R_{max}, above which it stabilizes. This apparently reflects once again the transition across the chemical divide. Closer examination of the m/z 216 mass chromatograms reveals that 2-methylpyrene (Peak 23) is favored over the other two methylpyrenes (Peaks 25 and 26) at high rank. This may be explained by the lesser steric hindrance at the C-2 position. An additional ratio could be created based on this phenomenon. Further rank effects are readily seen among the "Π_2" compounds (dimethylpyrenes and isomers) on m/z 230 traces.

The "X_0" tetraaromatic compounds also exhibit strong rank dependent effects (m/z 228, Figure 3). A sharp decrease in benzo[a]anthracene (Peak 27) concentrations relative to chrysene (Peak 28) is apparent. When the corresponding ratio ("CHR", Figure 4) is computed, the resulting curve shows a pronounced increase in slope between the 0.82 and 1.02% R_{max} vitrinites, as did the MPY ratio. Thus, the greatest loss of benzo[a]anthracene occurs during the transition corresponding to the chemical divide and the Teichmüller coalification jump (1). Distributions of the methylated variants of these compounds ("X_1"), as observed on m/z 242 mass chromatograms, are also significantly affected by the rank increase between 0.82 and 1.02% R_{max}.

Conclusions

Through the careful use of multi-step density gradient centrifugation, vitrinite concentrates of high purity were isolated from coal of the Lower Kittanning seam rank series. Chemical analyses of these concentrates permit one to focus on rank effects without concern for variations in organic matter type, over coal ranks corresponding to vitrinite reflectances ranging from 0.66 to 1.39% R_{max}, that is, the entire oil generation window.

In addition to the previously recognized losses of phenolic compounds with increasing rank, the vitrinite pyrolyzates exhibit marked increases in relative and absolute concentrations of tri- and tetraaromatic hydrocarbons, especially methylphenanthrene, methylfluorene, benzo[a]fluorene and methylchrysene isomers. There are only minor changes in the concentrations of mono- and diaromatic hydrocarbons — they are significant components at all rank levels examined. The 3- and 4-ring compounds are likely the products of the condensation of phenolic structures upon reduction.

The distributions of individual isomers of many of the polyaromatic hydrocarbons in the pyrolyzates also show pronounced rank effects. These phenomena may be exploited by the creation of ratios which respond to increases in the level of thermal alteration. Two such ratios are particularly effective over the full rank range studied. One depends on the thermal stability of the 1,3,7-, 1,3,6- and 2,3,6-trimethylnaphthalenes and the other, on the relatively robust 2,6- and 2,7-dimethylphenanthrenes. Other ratios exploit the thermal instability of anthracene and methylanthracene, compounds readily detectable in the pyrolyzates of low rank vitrinite.

Two other thermal indicator ratios computed for the pyrolyzates show sharp increases between the 0.82 and 1.02% R_{max} vitrinites. These ratios are based upon the superior stability of chrysene over benzo[a]anthracene and of benzo[a]fluorene over methylpyrenes. This is coincident with the sudden overall increase in tri- and tetraaromatic hydrocarbon concentrations. This "chemical divide" occurs at the same rank level as the   second coalification jump of Teichmüller (1). Thus, petrographically-recognizable physical transformations in coal are shown be coincident with major chemical restructuring of the vitrinite.

Vitrinite has long been employed petrographically as a rank indicator. It now appears that its usefulness may be extended into the chemical realm, particularly through the use of flash pyrolysis. The isolation of high purity vitrinite concentrates permits analysis with the least possible interference by other organic matter types. However, as a practical matter, pyrolysis of organic matter samples with naturally high concentrations of vitrinite should also produce suitable results.

Literature Cited

1. Teichmüller, M. *Org. Geochem.* **1986,** *10,* 581-599.
2. Shinn, J. H. *Fuel* **1984,** *63,* 1187-1196.
3. Carlson, G. A. *Energy & Fuels* **1992,** *6,* 771-778.
4. Nishioka, M.; Larsen, J. W. *Energy & Fuels* **1990,** *4,* 100-106.
5. Hatcher, P. G.; Faulon, J.-L.; Wenzel, K. A.; Cody, G. D. *Energy & Fuels* **1992,** *6,* 813-820.
6. Given, P. H.; Marzec, A. *Fuel* **1988,** *67,* 242-244.
7. Haenel, M. W. *Fuel* **1992,** *71,* 1211-1223.
8. Nip, M.; de Leeuw, J.W.; Schenk, P.A. *Geochim. Cosmochim. Acta* **1988,** *52,* 637-648.
9. Nip, M.; de Leeuw, J.W.; Schenk, P.A.; Windig, W.; Meuzelaar, H.L.C.; Crelling, J.C. *Geochim. Cosmochim. Acta* **1989,** *53,* 671-683.
10. Giuliani, J. D.; Wang, P.; Dyrkacz, G. R.; Johns, R. B. *J. Anal. Applied Pyrolysis* **1991,** 20, 151-159.
11. Nip, M.; de Leeuw, J.W.; Crelling, J.C. *Energy & Fuels* **1992,** *6,* 125-136.
12. Stankiewicz, B. A.; Kruge, M. A.; Crelling, J. C. In *Programme et Recueil des Résumés, IXe Colloque International des Pétrographes Organiciens Francophones, 16-18 Juin 1993*; Elf Aquitaine Production: Pau, France.
13. Kruge , M. A.; Stankiewicz, B. A.; Crelling, J. C. In *Organic Geochemistry: Poster Sessions from the 16th International Meeting on Organic Geochemistry,*

Stavanger, 1993; Øygard, K., Ed.; Falch Hurtigtrykk: Oslo, Norway, 1993; pp. 140-144.

14. Tissot, B. P.; Welte, D. H. *Petroleum Formation and Occurrence*; Springer-Verlag: Berlin, 1984; pp. 151-158 and 234-241.

15. Hatcher, P. G.; Lerch, H. E.; Bates, A. L.; Verheyen, T. V. *Org. Geochem.* **1989**, *14*, 145-155.

16. Hatcher, P. G. *Org. Geochem.* **1990**, *16*, 959-968.

17. Crelling, J.C. *Am. Chem. Soc. Div. Fuel Chem. Prepr.* **1989**, *34 (1)*, 249-255.

18. Hower, J. C.; Davis, A. *Geol. Soc. Am. Bull.* **1981**, 92, 350-366.

19. Houseknecht, D. W.; Bensley, D. F.; Hathon, L. A.; Kastens, L. A. *Org. Geochem.* **1993**, 20, 187-196.

20. Bensley, D. F.; Crelling, J. C. *Proceedings of the Int. Conf. on Coal Science*, 1993, Vol. 1, pp. 578-581.

21. Radke, M.; Welte, D.H. In *Advances in Organic Geochemistry 1981*; Bjøroy M. et al., Eds.; John Wiley & Sons, Ltd.: Chichester, 1983; pp. 504-512.

22. Sentfle, J. T.; Larter, S. R.; Bromley, B. W.; Brown, J. H. *Org. Geochem.* **1986**, 9, 345-350.

23. Venkatesan, M. I.; Ohta, K.; Stout, S. A.; Steinberg, S.; Oudin, J. L. *Org. Geochem.* **1993**, 20, 463-473.

24. Solum, M. S.; Pugmire, R. J.; Grant, D. M. *Energy & Fuels* **1989**, *3*, 187-193.

25. Radke, M. In *Advances in Petroleum Chemistry, Vol. 2*; Brooks J.; Welte D. H., Eds.; Academic Press: London, 1987, pp. 141-207.

26. Kruge, M. A.; Landais, P. *Preprints of Papers Presented at the 204th ACS National Meeting*; Amer. Chem. Soc. Div. of Fuel Chemistry: Washington, DC, 1992, Vol. 37, No. 4, pp. 1595-1600.

27. Requejo, A. G.; Gray, N. R.; Freund, H.; Thomann, H.; Melchior, M. T.; Gebhard, L. A.; Bernardo, M.; Pictroski, C. F.; Hsu, C. S. *Energy & Fuels* **1992**, *6*, 203-214.

28. Alexander, R.; Kagi, R.I.; Rowland, S.J.; Sheppard, P.N.; Chirila, T.V. *Geochim. Cosmochim. Acta* **1985**, *49*, 385-395.

29. Radke, M.; Welte, D.H.; Willsch, H. *Org. Geochem.* **1986**, *10*, 51-63.

30. Budzinski, H.; Garrigues, P.; Radke, M.; Connan, J.; Rayez, J. C.; Rayez, M. T. *Energy & Fuels* **1993**, *7*, 505-511.

31. Garrigues, P.; Oudin, J.L.; Parlanti, E.; Monin, J.C.; Robcis, S.; Bellocq, J. *Org. Geochem.* **1990**, *16*, 167-173.

32. Budzinski, H.; Garrigues, P.; Radke, M.; Connan, J.; Oudin, J. L. *Org. Geochem.* **1993**, *20*, 917-926.

RECEIVED April 15, 1994

Chapter 10

Molecular Characterization of Vitrinite Maturation as Revealed by Flash Pyrolysis Methods

H. Veld[1,2], J. W. de Leeuw[3], J. S. Sinninghe Damsté[3], and W. J. J. Fermont[2]

[1]Laboratory of Paleobotany and Palynology, University of Utrecht, Heidelberglaan 2, 3584 CS Utrecht, Netherlands
[2]Geological Survey, P.O. Box 126, 6400 AC Heerlen, Netherlands
[3]Netherlands Institute for Sea Research (NIOZ), Division of Marine Biogeochemistry, P.O. Box 59, 1790 AB Den Burg, Netherlands

Curie-point pyrolysis - gas chromatography analyses were performed on four vitrinite concentrates of maturity levels from 0.96 %Rmax to 1.93 %Rmax. Analyses were carried out on "thermal extracts" obtained at a Curie-temperature of 358 °C and on pyrolysates obtained at a Curie-temperature of 770 °C. The relative concentrations of alkylphenols and alkylnaphthalenes decrease with increasing maturity. Alkylbenzenes are prominent pyrolysis products throughout the maturity interval investigated. A significant increase in the relative concentration of alkylphenanthrenes and alkylbiphenyls with increasing maturity is noted. There is a substantial similarity between the compositions of "thermal extracts" and pyrolysates. This indicates that the trends observed are mainly due to changes in frequency distributions of extractable compounds. Vitrinite reflectance measurements carried out on the unextracted and extracted concentrates revealed no significant differences with one exception. For one sample, with a maturity approximately at the transition from the "oil window" to the "gas window", solvent extraction resulted in a significantly higher reflectance value of the residue. This sample was rich in extractable n-alkanes.

A major constituent of many coals is vitrinite. The macerals of the vitrinite group are derived from plant cell wall material (woody tissue) by processes of humification. During the subsequent physico-chemical stage of coalification time and temperature are the main factors governing the transformation of organic matter. A wide variety of methods has been applied to monitor these thermally induced chemical and physical changes. The most important physical parameter to measure the degree of coalification or maturation is based on the progressive change in the capacity of macerals to reflect incident light. The

0097–6156/94/0570–0149$08.00/0
Published 1994 American Chemical Society

advantage of reflectance measurements is that the relatively small measurement area allows determinations to be carried out on specific macerals. Vitrinite reflectance is by definition measured on one vitrinite maceral, i.e. telocollinite (*1, 2*). Unlike most of the other macerals telocollinite shows a relatively gradual change with increasing coalification. Many of the chemical rank parameters are obtained from bulk samples of coal. There is evidence that the values of these parameters depend on the relative proportions of the different macerals present in the coal (*3, 4*). Previous studies demonstrated that flash pyrolysis methods are very useful for the characterization of the insoluble fraction of coals, coal macerals and their precursors on a molecular level (*5 - 9*). However, these studies concentrated either on the characterization of relatively low rank coals or did not account for specific coal petrographic compositions of the studied material. To date only a few studies applied analytical pyrolysis methods on high rank coals (*7, 10*). The aim of the present study is to examine in detail the molecular changes that take place in a series of high rank vitrinite macerals as revealed by flash pyrolysis.

Material

Four vitrinite concentrates were obtained from the European Centre for Coal Specimen (SBN). The concentrates were prepared from Upper Carboniferous (Westphalian) vitrinite-rich coals from the Ruhr area in Germany by means of chemical and mechanical comminution and density flow separation techniques (*11*). The density separations were carried out using mineral density solutions to avoid contamination of organic solvents. The purity of the concentrates ranges from 93.8 to 98.0 %. The vitrinite reflectance values of the concentrates range from 0.96 to 1.93 %Rmax, which classifies the coals as high volatile bituminous A to low volatile bituminous coals. A data summary is given in Table I. More detailed information on the chemical and petrographical properties of the parent coals and of the vitrinite concentrates is given elsewhere (*12*).

Table I. Data summary of the vitrinite concentrates

Code	Parent coals					Vitrinite concentrates					
	%Ro	SD	V	L	I	M	%Ro	SD	V	L	I	M
312	1.07	0.04	48.2	7.4	29.4	15.0	1.02	0.05	93.8	2.2	4.0	0.0
313	1.31	0.04	69.0	0.0	10.0	21.0	1.26	0.06	98.0	0.0	2.0	0.0
314	1.63	0.06	82.4	0.0	15.0	1.6	1.64	0.10	95.8	0.4	3.8	0.0
315	1.89	0.06	65.2	0.0	32.8	2.0	1.91	0.12	96.0	0.2	3.6	0.2

%Ro = mean random vitrinite reflectance, *SD* = standard deviation, *V* = %Vitrinite, *L* = %Liptinite, *I* = %Inertinite, *M* = %Mineral matter (Data provided by the European Centre for Coal Specimen).

Methods

Vitrinite reflectance. Mean maximum vitrinite reflectance measurements (%Rmax) were carried out following standard procedures (*1, 2*). Vitrinite reflectance was measured on polished blocks of the parent coals, the unextracted and the extracted vitrinite concentrates. The different fractions of the samples have the extensions P, B and R, respectively (see Table II). Measurements were carried out under oil immersion at a wavelength of 546 nm using a Leitz MPV III microscope system.

Curie point pyrolysis - gas chromatography (Py-GC) and Curie point pyrolysis-gas chromatography - mass spectrometry (Py-GC-MS). The Py-GC analyses were performed on a Varian 3700 gas chromatograph, equipped with a FOM-3LX unit for pyrolysis (*8*). The samples were applied to a ferromagnetic wire (*13*). A Curie point high-frequency generator (Fischer, Model 9425) was used to generate the magnetic field. The ferromagnetic wires were inductively heated in 0.15 s to their Curie temperatures, 770 °C for pyrolysis and 358 °C for "thermal extraction", and were kept at their final temperature for 10 s. Separation of the pyrolysis products was achieved using a fused-silica column (25 m x 0.32 mm I.D.) coated with CP Sil-5 (film thickness 0.45 μm). Helium was used as the carrier gas. The oven was programmed from 0 °C (5 min isothermal) to 320 °C (10 min isothermal) at a rate of 3 °C/min. Ultrasonic solvent extraction of the vitrinite concentrates was executed using methanol-dichloromethane (1:1). Thermal extraction (358 ° C) as well as flash pyrolysis (770 °C) was performed on the unextracted and the solvent extracted vitrinite concentrates.

For the Py-GC-MS analyses, the same pyrolysis unit, capillary columns, carrier gas and temperature conditions of the HP 5890 Hewlett Packard gas chromatograph were used, as mentioned for the Py-GC analyses. The GC column was directly inserted into the ion source of a VG-70S mass spectrometer. Electron impact spectra were obtained at 70 eV using the following conditions: mass range m/z 50-900; cycle time 1.8 s.

Results

The results of the reflectance measurements are presented in Table II as the mean of 50 readings and their standard deviation. Except for sample 313 no significant differences in reflectance values were recorded between the three different sample fractions. For sample 313 the measured reflectance of the parent coal and its vitrinite concentrate are almost identical whereas the reflectance value of the extracted concentrate is 0.16 %Rmax higher, which is considered significant.

In Figures 1a-d the chromatograms of the pyrolysates of the un-extracted samples are shown. The identification of the components released was by Py-GC-MS and is given in Table III.

A considerable decrease in the number of components with increasing rank, especially between the samples 313 and 314, is noted. The relative proportions of the constituents also change with increasing rank. Figure 1a shows the partial Py-GC trace of sample 312-B. This pyrolysate is mainly characterized by the presence of alkylbenzenes, alkylphenols, alkylnaphthalenes and, to a lesser degree, by homologous series of n-alkanes and n-alk-1-enes. Sample 313-B has a similar composition although the relative abundances of the individual components differ to some extent. Both the C_2-alkylated naphthalenes as well as n-alkanes are the most dominant peaks in the pyrolysate of this sample.

Table II. Vitrinite reflectance measurements

Code	P		B		R	
	%Rmax	SD	%Rmax	SD	%Rmax	SD
312	0.98	0.06	0.96	0.04	0.96	0.05
313	1.23	0.05	1.21	0.06	1.37	0.04
314	1.66	0.05	1.59	0.07	1.64	0.05
315	1.93	0.06	1.93	0.07	1.88	0.06

P = Parent coal; B = Unextracted vitrinite concentrate; R = Extracted vitrinite concentrate; SD = Standard deviation

The pyrolysates of samples 314-B and 315-B are dominated by alkylbenzenes, alkylnaphthalenes, alkylanthracenes and alkylbiphenyls. Straight chain n-alkanes, n-alk-1-enes and alkylphenols are still present as minor constituents and differences exist in the relative proportions of the individual components. As can be deduced from Figure 1 there is a distinct maximum in the relative abundance of the n-alkanes and the C_2- and C_3-alkylated naphthalenes (Q and T) in the pyrolysate of sample 313-B, as compared to the other samples.

The chromatograms of the flash pyrolysates of the extracted vitrinite concentrates (Figures 2a-d) and their unextracted counterparts (Figures 1a-d) reveal only minor differences in composition. The main difference is a significant reduction of the relative abundance of the C_2-alkylated naphthalenes and the n-alkanes. This is most obvious in sample 313 (Figures 1b and 2b).

The chromatograms of the series of "thermally extracted" vitrinite concentrates which were previously solvent extracted show a very high signal to noise ratio due to the relative efficient removal of soluble compounds. This series is not further discussed here.

The results obtained by the "thermal extraction" at a Curie temperature of 358 °C show that the thermal extracts have a similar composition as the pyrolysates of the unextracted vitrinite concentrates (Figure 3). The proportions, however, may differ significantly. The main differences in comparison to the pyrolysates are a clear dominance of alkylnaphthalenes and a lower yield of alkylphenols and alkylbenzenes. n-Alkanes are present as

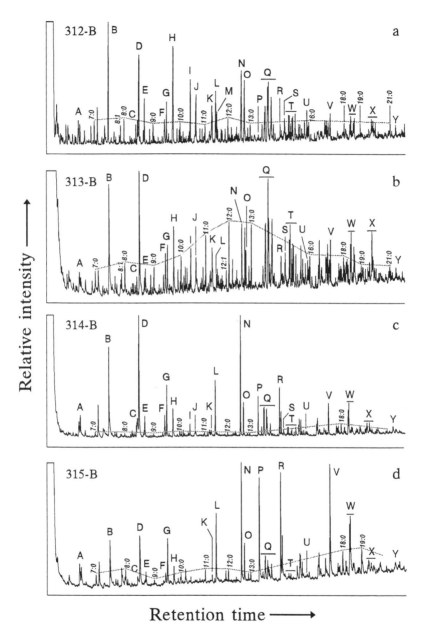

Figure 1. Partial chromatograms of the flash pyrolysates (Curie temperature 770 °C) of the unextracted vitrinite concentrates. Peak identifications are indicated in Table III. #:1 = n-alk-1-enes, #:0 = n-alkanes. Vitrinite reflectance values (%Rmax) of each sample is given in Table II. For analytical conditions, see Methods section.

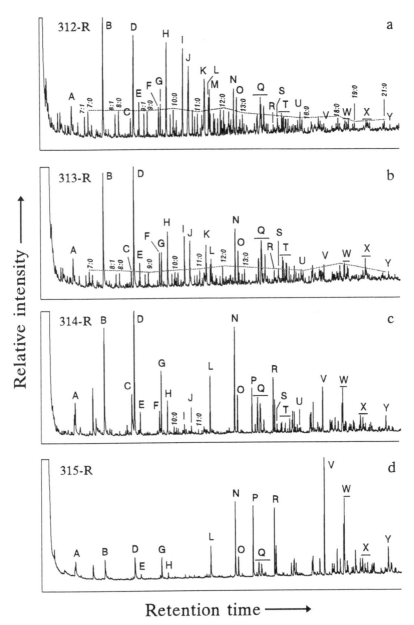

Figure 2.   Partial chromatograms of the flash pyrolysates (Curie temperature 770 °C) of the extracted vitrinite concentrates. Peak identifications are indicated in Table II. #:1 = n-alk-1-enes, #:0 = n-alkanes. Vitrinite reflectance values (%Rmax) of each sample is given in Table II. For analytical conditions, see Methods section.

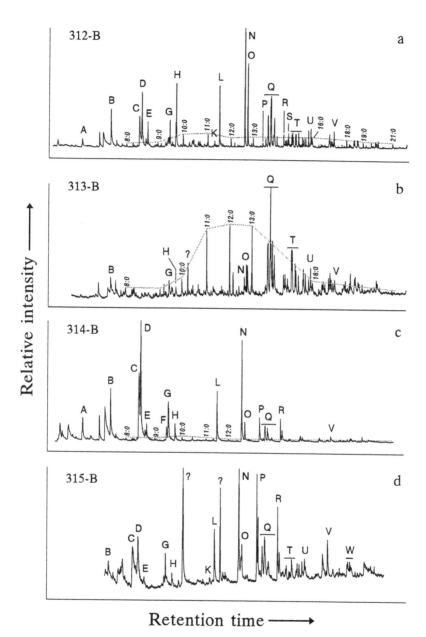

Figure 3. Partial chromatograms of the "thermal extracts" (Curie temperature 358 °C) of the unextracted vitrinite concentrates. Peak identifications are indicated in Table III. #:1 = n-alk-1-enes, #:0 = n-alkanes. For analytical conditions, see Methods section.

minor components in the thermal extracts, the n-alk-1-enes are virtually absent. As an exception the thermal extract of sample 313 is dominated by n-alkanes (Figure 3b).

A semi-quantitative analysis of the four samples is presented in Figure 4. For this figure the peak areas of most of the aromatic components identified in Table III have been calculated for both the unextracted and the extracted vitrinite concentrates. The relative proportions of the different components are normalized to the largest peak in the chromatogram. Although C_2-alkylated naphthalenes are the most dominant components in the pyrolysates of the samples 312 and 313, solvent extraction results in a distinct relative increase of alkylbenzenes and alkylphenols. The differences between the pyrolysates of the unextracted and the extracted vitrinites of the samples 314 and 315 are less distinct.

In Figure 5 a summary of the relative abundance of four classes of aromatic components present in the pyrolysates is presented as a function of maturity (vitrinite reflectance). These groups represent a summation of the components given in Table III. The most obvious trend in both the unextracted and extracted vitrinites with increasing maturity is the relative decrease of phenols. In both fractions of all samples benzenes generally constitute over 30 % of the pyrolysis products quantified. The total sum of the naphthalenes also remains high between 25 to 40 %. Furthermore, the relative abundance of phenanthrenes shows a significant increase with increasing maturity. Although not shown here biphenyls show an almost identical trend as the phenanthrenes.

Table III. Major components identified in the flash pyrolysates and the thermal extracts

	Component		Component
A	Benzene	M	3,5-Dimethylphenol and 3-Ethylphenol
B	Toluene		
C	Ethylbenzene	N	2-Methylnaphthalene
D	1,3 and 1,4-Dimethylbenzene	O	1-Methylnaphthalene
E	1,2-Dimethylbenzene	P	Biphenyl
F	1-Methyl-3-ethylbenzene and 1-Methyl-4-ethylbenzene	Q	C_2-Naphthalenes
		R	C_1-Biphenyls
G	1,3,5-Trimethylbenzene	S	Dibenzofuran
H	Phenol (and 1,2,4-Trimethylbenzene)	T	C_3-Naphthalenes
		U	C_1-Dibenzofurans
I	2-Methylphenol	V	Phenanthrene
J	3 and 4-Methylphenol	W	C_1-Phenanthrenes/Anthracenes
K	2,4-Dimethylphenol	X	C_2-Phenanthrenes/Anthracenes
L	Naphthalene	Y	C_1-Pyrenes

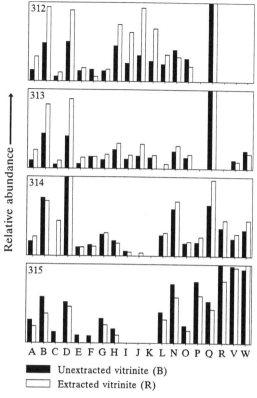

Figure 4. Semi-quantitative representation of the concentrations of the components identified in Table III for the unextracted and extracted vitrinite concentrates. Peak areas were calculated relative to the largest peak in each chromatogram.

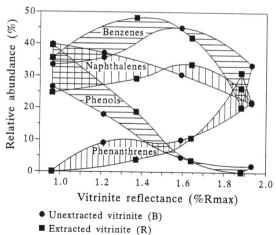

Figure 5. Schematic representation of the relative concentrations of the aromatic components as a function of maturity.

Discussion

The wide maturity range and low number of samples do not allow for a fine-tuning of the trends observed. The observed differences between samples of different rank have an exclusively semi-quantitative aspect because all the pyrolysis products identified at a reflectance value of 0.96 %Rmax are also present in the sample with a reflectance value of 1.93 %Rmax. The reduction in the yield of alkylphenols upon pyrolysis with increasing maturity has been discussed in earlier papers (7, 14). A similar result is obtained here. The relative abundance of the alkylbenzenes shows a rather wide scatter throughout the coalification range investigated and appears to be little affected by increasing coalification. The abundance of the alkylphenanthrenes and alkylbiphenyls clearly increases with increasing maturity. The increase in the relative yields of the n-alkanes and the C_2- and C_3-alkylnaphthalenes followed by a decrease is likely to be related to hydrocarbon generation and subsequent expulsion from the vitrinite matrix as coalification continues. The reflectance value of 1.21 %Rmax approximates the end of the bituminization range of coals which is set at a reflectance value of approximately 1.30 %Rmax (15). This reflectance value coincides with the transition from the oil window to the gas window and with one of the "coalification jumps" of liptinites (15). Because pure vitrinite concentrates were used in the present study the differential effects of maceral composition upon the character of the pyrolysis yields can be neglected. Moreover, because of their very low abundance in the parent coals, liptinite macerals are not regarded here as a possible source for the n-alkanes and alkylnaphthalenes. However, the possibility of different source materials for the vitrinites can not be excluded. The most important biopolymeric precursor of vitrinite is considered to be lignin. The different pyrolysis products of extant lignin are well-documented (e.g. 6) and the transformation of lignin up to the subbituminous stage of coalification (0.5 - 0.6 %Rmax) is relatively well-established (16, 17), because the coalified tissue up to this maturity level can be morphologically related to woody cell tissue (xylem). With increasing coalification such morphological features gradually disappear. Moreover, because of the intimate association of pure telinite (cell tissue) with collinites (e.g. cell fillings) in bituminous coals the differentiation between both is extremely difficult. It has recently been demonstrated that the precursor of collinites may consist of highly aliphatic constituents which have migrated into the cell lumina (18, 19). Upon pyrolysis these materials yield homologous series of n-alkanes and n-alk-1-enes. Benzenes, naphthalenes, and phenanthrenes are not reported to be pyrolysis products of wood lignins (6). This suggests that these products are formed as a result of increasing coalification. However, recent studies on gymnospermous bark tissues, which contain lignin moieties, indicate that benzenes and naphthalenes may also be pyrolysis products of extant lignins (20). It is suggested (e.g. 21) that incorporation of other components in the vitrinite matrix during early degradation (e.g. bacterial or fungal-derived) and/or coalification (impregnation by liquid hydrocarbons) is not unlikely.

The above findings support the view that the "vitrinite macromolecule" may show considerable chemical variation inherited from precursors which are not exclusively lignins. This chemical variation will to some extent also be recorded by vitrinite reflectance measurements. The results of sample 312 reveal that at identical reflectance values the chemistry of a sample may show considerable, mainly semi-quantitative, differences. This indicates that the extractable fraction has no significant influence on vitrinite reflectance values. However, for sample 313 a significant difference of 0.16 %Rmax has been recorded between the reflectance values of the unextracted and the extracted vitrinite concentrate. This sample, unlike the other three samples, is dominated by (extractable) C_2-alkylated napthalenes and n-alkanes. It thus appears that there is a relationship between the presence of these components and the vitrinite reflectance value of this sample. In several studies suppressed vitrinite reflectance values are related to higher hydrogen contents of the vitrinite (21-24). This is supported by the considerable higher aliphatic character of this sample as revealed by flash pyrolysis. From our study it may be concluded that the extractable fraction of aliphatic compounds present in the vitrinite matrix is capable to absorb a significant fraction of photons from an incident light beam. As a consequence the removal of this fraction results in a higher vitrinite reflectance value for the extracted vitrinite concentrate. The results are not conclusive whether this chemical differentiation is the result of early diagenetic variation or caused by hydrocarbon generation within the vitrinite maceral. Because of its position within the "oil window" it may be a combination of both factors (25).

Acknowledgments

This work was supported by the Netherlands Foundation for Earth Science Research (AWON) with financial aid from the Netherlands Organization for Scientific Research (NWO). We gratefully acknowledge the European Centre for Coal Specimen (SBN, Eygelshoven, the Netherlands) for providing the samples. Dr. E. Tegelaar is thanked for critically reviewing the manuscript. This is contribution no. 323 of the Division of Marine Biogeochemistry.

Literature cited

1   International Committee of Coal Petrology. *International Handbook of Coal Petrography*; CNRS: Paris, 1963; 2nd. edn.
2   International Committee of Coal Petrology. *International Handbook of Coal Petrography*; CNRS: Paris, 1971, 1975; Suppl. to 2nd. edn.
3   *Stach's Textbook of Coal Petrology*; Stach, E., Mackowsky, M., Teichmüller, M., Taylor, G.H., Chandra, D. & Teichmüller, R., Eds.;   Gebrüder Borntraeger: Berlin, 1982.
4   Radke, M.; Welte, D.H.; Willsch, H. *Org. Geochem.* **1986**, *10*, pp. 51-63.
5   Schenck, P.A.; de Leeuw, J.W.; Viets, T.C.; Haverkamp, J. In *Petroleum Geochemistry and Exploration of Europe;* Brooks, J., Ed.;   Geol. Soc. Spec. Publ.; Blackwell Scientific Publications: London, 1983, Vol. 12; pp. 267-274.

6   Saiz-Jiminez, C.; de Leeuw, J.W. *Org. Geochem.* **1986**, 10, pp. 869-876.
7   Senftle, J.T.; Larter, S.R.; Bromley, B.W.; Brown, J.H. *Org. Geochem.* **1986**, 9, pp. 345-350.
8   Boon, J.J.; Pouwels, A.D.; Eijkel, G.B. *Biochem. Soc. Trans.* **1987**, 15, pp. 170-174.
9   Nip, M.; de Leeuw, J.W.; Crelling, J.C. *Energy & Fuels*, **1992**, 6(2) pp. 125-136.
10 Tromp, P.J.J.; Moulijn, J.A.; Boon, J.J. In *New Trends in Coal Science;* Yürüm, Y., Ed.; Kluwer Academic Publishers: Dordrecht, 1988; pp. 241-269.
11 Fermont, W.J.J.; Joziasse, J.; Nater, K.A.; van der Veen, A.H.M. In *Coal Science and Technology and Related Processes, 1993. Third International Rolduc Symposium on Coal Scienc;* Kapteijn, F.; Moulijn, J.A.; Nater, K.A., Prins, W. Eds.; Spec. Issue Fuel Processing and Technology; Elsevier: Amsterdam, 1993, Vol. 36; pp. 33-39.
12 Fermont, W.J.J.; Joziasse, J.; Nater, K.A.; van der Veen, A.H.M. In *Coal Science and Technology and Related Processes, 1993. Third International Rolduc Symposium on Coal Scienc;* Kapteijn, F.; Moulijn, J.A.; Nater, K.A., Prins, W. Eds.; Spec. Issue Fuel Processing and Technology; Elsevier: Amsterdam, 1993, Vol. 36; pp. 41-46.
13 Venema, A.; Veurink, J. *J. Anal. Appl. Pyrolysis* **1985**, 7, pp. 207-213.
14 Larter, S. *Geol. Rundschau* **1989**., 78(1), pp. 349-359.
15 Teichmüller, M. In *Coal and Coal-bearing Strata: Recent Advances;* Scott, A.C. Ed.; Geol. Soc. Spec. Pub.; Blackwell Scientific Publications: London, 1987, Vol. 32; pp. 127-169.
16 Hatcher, P.G.; Lerch, H.E.; Verheyen, T.V. *Int. J. Coal Geol.*, **1989a**, 13, pp. 65-97.
17 Hatcher, P.G.; Wilson, M.A.; Vassallo, A.M.; Lerch, H.E. *Int. J. Coal Geol.*, **1989b**, 13, pp. 99-126.
18 Zhang, E.; Hatcher, P.G.; Davis, A. *Org. Geochem.*, **1993**, 20, pp. 721-734.
19 Tegelaar, E.W.; de Leeuw, J.W.; Saiz-Jiminez, C. *Sci. Total Eviron.*, **1989**, 81/82, pp. 1-17.
20 Tegelaar, E.W.; Hollman, G.; van der Vegt, P.; de Leeuw, J.W.; Holloway, P.J. *Org. Geochem.*, **submitted**
21 Powell, T.G.; Boreham, C.J.; Smyth, M.; Russell, M.; Cook, A.C. *Org. Geochem.*, **1991**, 17, pp. 375-394.
22 Price, L.C.; Barker, C.E. *J. Petr. Geol.*, **1985**, 8(1), pp. 59-84.
23 Wenger, L.M.; Baker, D.R. *Org.Geochem.*, **1987**, 11, pp. 411-416.
24 Hutton, A.C.; Cook, A.C. *Fuel*, **1980**, 59, pp. 711-714.
25 Collinson, M.E.; van Bergen, P.; Scott, A.; de Leeuw, J.W. In *Coal and Coal-bearing strata;* Scott, A.; Fleet, A.J., Eds., **in press**.

RECEIVED June 3, 1994

Chapter 11

Influence of Pressure on Pyrolysis of Coal

R. J. Hill[1], P. D. Jenden[2], Y. C. Tang[2], S. C. Teerman[2],
and I. R. Kaplan[1]

[1]Department of Earth and Space Sciences, University of California,
Los Angeles, CA 90024
[2]Chevron Petroleum Technology Company, La Habra, CA 90633

The influence of pressure on the gas, liquid and solid products of coal pyrolysis has been investigated. The starting coal was Illinois #6, a high volatile bituminous coal (R_o=.5%) obtained from the Argonne premium coal collection. Dry, confined pyrolysis was performed in sealed gold tubes at 300 ^{o}C and 340 ^{o}C and pressures ranging from 70 to 2000 bars for 72 hours. Results show the rates of gas and liquid product generation and solid maturation are influenced by pressure. At 300 ^{o}C and 340 ^{o}C, rates of vitrinite maturation and CH_4 and CO_2 gas generation increase with pressure to about 600 bars (R_o=1.01% and 1.56%, respectively) and then decrease with further increase in pressure to 2000 bars (R_o=.83% and 1.43%, respectively). The opposite trend is observed in the pristane/n-C17 ratio with pristane/n-C17 reaching a minimum at 600 bars. The carbon isotope fractionation observed in methane and CO_2 due to changing pressure is less than .2 $^o/_{oo}$. These results suggest that although the effect of pressure on organic maturation is secondary, neglecting it could lead to errors in the interpretation of maturity based on vitrinite reflectance and influence the modeling of hydrocarbon generation in a basin.

The maturation of organic matter leading to the generation of oil and gas has traditionally been regarded as a temperature controlled process (1) with pressure being of subordinate importance (2). More recent work suggests pressure may have an important effect on the maturation of organic matter, but conflicting results have been reported by different authors. Some reported retardation of hydrocarbon generation by pressure, others enhancement of generation while still others report no pressure effect at all. This may, in part, be due to the different methods and experimental conditions used to study the influence of pressure on organic maturation.

0097–6156/94/0570–0161$10.34/0

Hesp and Rigby (3) demonstrated retardation of the thermal cracking of Gippsland basin oil with increasing pressure. McTavish (4) reported a suppression of vitrinite reflectance in North Sea samples with increasing burial pressure, suggesting pressure retarded organic maturation. Cecil et al. (5) found the coalification process to be retarded by increased pressure in autoclave experiments. Horvath (6) investigated the influence of pressure on the coalification process. The role of vapor pressure was investigated using a bomb where volatiles generated during pyrolysis accumulated in the bomb (closed system), providing a pressure of 60 bars during the experiment, while the role of load pressure was investigated using a bomb where pressure was applied using pistons and the volatiles produced during pyrolysis were allowed to escape (open system). Horvarth found mechanical pressure did not increase coal rank during pyrolysis for experiments performed between 90° C and 200° C and pressures of 1,500 bars and 10,000 bars in the open system while pyrolysate vapor pressure in the closed system slightly retarded coalification. Sajgo, et al. (7) using the methods of Horvarth (6) and temperatures between 200° C and 450° C found increasing system pressure retards hydrocarbon generation and coal maturation. Sajgo et al. only used three pressures in their study, 60 bars, 1000 bars and 2,500 bars. Both Sajgo et al. (7) and Horvarth (6) did not investigate pressures where most oil and gas is generated in nature. Price and Wegner (8) performed hydrous pyrolysis experiments in stainless steel bombs using whole-rock chips at temperatures between 175° C and 450° C and monitored overpressuring (up to 1,100 bars) in the bombs after the experiments were finished. They concluded that high hydrostatic fluid pressures retard all aspects of hydrocarbon generation, maturation and thermal destruction. Based on their experimental results, as well as the existence of measurable to moderate quantities of $C_{15}+$ hydrocarbons in ultra-high rank rocks (7; 9-13), Price and Wegner suggest all aspects of hydrocarbon generation are controlled by pressure. It is unclear, however, what the actual pressure was during their experiments. Blanc and Connan (14) found in sealed gold tube pyrolysis experiments performed at 330° C and pressures of 170, 550 and 1,000 bars that pressure retards the generation of hydrocarbons from organic matter, although an unexplained enhancement of generation is suspected between 550 bars and 1,000 bars. Blanc and Connan also found the expulsion efficiency of $C_{13}+$ hydrocarbons increases with pressure. Perlovsky and Vinkovetsky (15) argued theoretically that the positive activation volume of cracking reactions should cause pressure to retard the organic maturation process. Domine (16) found olefins to decrease and heavy hydrocarbon products (C_6+) to increase as pressure increased during the pyrolysis of hexane. While the body of data supporting the role of pressure in the retardation of organic maturation is growing, other research has produced conflicting results.

Monthioux et al. (17) performed sealed gold tube pyrolysis of coal at temperatures between 250° C and 550° C and pressures of 500 bars, 1,000 bars, 3,000 bars and 4,000 bars. No significant pressure effect on the liquid pyrolysis products was reported. Monthioux et al. (18) concluded that natural maturation is better simulated under confined conditions based on results from sealed gold tube pyrolysis of coal performed at temperatures between 250° C and 400° C and pressures of 500 bars and 1,000 bars, but did not report a pressure influence on maturation. Monthioux et al., however, did not measure gas yields or vitrinite reflectance and only investigated 2 pressures normally observed in sedimentary basins.

Braun and Burnham (*19*) concluded increasing pressure should enhance hydrocarbon thermal cracking up to about 400 bars before retarding generation, although they do not show retardation in their modeled results. Burnham and Braun based their conclusions on literature results (*20;21*). These conflicting results and conclusions, the variety of methods and conditions utilized and the lack of work concentrating on pressures normally observed in sedimentary basins have prompted us to further investigate pressure influences on organic maturation. We have utilized the confined pyrolysis technique using sealed gold tubes to examined the influence of increasing pressure on the maturation of coal. Temperatures were chosen that will produce enough liquid and gas hydrocarbons to evaluate the influence of pressure on organic maturation with eight pressures chosen between 70 bars and 2,000 bars, concentrating on pressures where oil and gas is generated.

Methods and Materials

The hydrothermal laboratory in the Department of Earth and Space Sciences at UCLA was utilized for sealed gold tube pyrolysis experiments. The laboratory and materials used in the experiments are described below.

Starting Material. The starting material for the confined system pyrolysis experiments was Illinois Basin Coal #6 from the Argonne National Laboratory premium coal collection. Illinois Basin Coal #6 is a high volatile bituminous coal (R_o=.5%) (*22,23*). The coal was stored in a brown glass bottle under argon at all times to prevent oxidation.

Gold Tube Preparation. The gold tubing used in the pyrolysis experiments measured 4 mm outer diameter and 3.62 mm inner diameter. The gold tubing was cut with a scalpel into 5.0 cm lengths and cleaned first by washing in boiling 6N HCl and then rinsing 3 times with dichloromethane and methanol, respectively. The gold tube was then annealed at 900° C for 8-12 hours.

Sample Loading and Gold Tube Welding. After annealing, one end of the 5.0 cm tubes was crimped using smooth faced needle nose pliers, clipped to produce a straight edge and then welded shut using an acetylene torch. The weld was then inspected for blemishes using a 6x microscope and rewelded if necessary. Approximately 100 mg of coal was then loaded into a gold tube in a glove box containing an argon atmosphere. After sample loading, the open end of the gold tube was crimped parallel to the first weld and weighed. The mass of coal in the gold tube was determined by difference. The unwelded edge of the gold tube was then clipped as described above and welded using an arc welder. During arc welding, the bottom of the gold tube was placed in a beaker of water and a drop of water is placed around the edge to be welded to keep temperature inside the gold tube to a minimum during welding. The sealed gold tube capsule was weighed again after the second welding for comparison with the capsule weight after the experiment.

Bombs. The bombs utilized in the hydrothermal laboratory are 31 cm long, have a diameter of 3.2 cm and a bore diameter of 7 mm, are constructed of Rene (nickel alloy) and have a steel cap. Water pressure (up to 3,000 bars) is applied through the cap via a 1/8" o.d. stainless steel line with a cone-in-cone fit to the bomb housing. The read thermocouple from the control unit is inserted into a port in the back of the bomb for temperature monitoring. Heavy Duty Electric Company (Milwaukee, WI) brand, 13" length x 6.5" diameter split furnaces are used in all experiments.

Temperature Control. Furnace temperature was controlled and monitored by a custom built 40 channel controller unit manufactured at UCLA. Each furnace was equipped with an American National Standards Institute Type K chromel/alumel (nickel-chromium versus nickel aluminum) thermocouple used by the controller unit to adjust and control the furnace temperature. Each furnace was also equipped with a Type K thermocouple for temperature read out which was monitored at the control unit. The type K thermocouple has a range of -50° C to 1200° C, an accuracy of $+/-2.2^\circ$ C or .75% (whichever is greater) and a readout resolution of 1.0° C. Temperature was read on an Omega Engineering Incorporated Model 115 digital readout with a resolution of 1° C. In order to minimize experimental error, temperatures for all furnaces were calibrated with a single thermocouple.

Pressure Control. Water pressure was manually controlled using a pressure pump (Owatonna Tool Company Model A Pressure Intensifier, Owatonna, MN). Pressure was monitored using a pressure gauge (Heisse, $+/-$ 0.1%) which can be read to +/- 1 bar.

Pyrolysis. Experiments were conducted for 72 hours, isothermally at 300° C and 340° C ($+/-2.5^\circ$ C) and pressures ranging from 70 bars to 2000 bars. No water was added. The experimental times and temperatures were chosen based on the work of Lewan (24) to approximate vitrinite maturation near the beginning and end of oil generation. One gold tube containing sample was placed in each bomb followed by a solid stainless steel rod (4 mm o.d.) which occupied essentially all the remaining void space. The bomb was then filled with water, the bomb cap screwed in place, and the bomb placed in a furnace. Ceramic spacers of constant length were used to insulate the end of each bomb and to ensure that the position of the bomb in the furnace was always the same. The read thermocouple was inserted through a hole in the ceramic spacer and into a port in the end of the bomb (Figure 1) for monitoring temperature at the controller unit (Figure 2). Temperature was raised incrementally over a period of 90 minutes such that the target temperature was never exceeded. Pressure was maintained near the pressure of interest, but not in excess, until the target temperature was reached. The pressure of interest was then set. Temperature and pressure were monitored at least twice daily, and adjusted as needed, until the experiments were completed.

After the experiments were complete, the bombs were cooled in an air stream for approximately ten minutes and the gold tubes then removed from the bombs, weighed, and the contents analyzed. If the weight of the gold tube after the experiment was within +/- 0.25 mg of the weight of the gold tube before the experiment, the run was considered successful.

Gas Product Analysis. Product analysis was performed at Chevron Petroleum Technology Company and included analysis of the non hydrocarbon and hydrocarbon gases, hydrocarbon liquids and residual solids. Gas collection was initiated by puncturing the gold tube under vacuum. The pyrolysate gas was initially exposed to the dry ice/acetone trap (T=-77° C) for 4-5 minutes. The gas sample was then expanded into the liquid nitrogen trap (T=-196° C) for 4-5 minutes. The liquid nitrogen non-condensable gases (N_2, CH_4, CO and H_2) were collected in the gas burette using a Toepler pump, the total fraction quantified, and introduced directly onto a gas chromatograph (GC). After GC analysis of the liquid nitrogen non-condensable gases had begun, the liquid nitrogen bath was removed and the CO_2 and C_2 -C_5 fraction was collected, quantified and frozen into a pyrex tube.

The liquid nitrogen non-condensable gases were analyzed with a two-channel United Technologies/Packard 438A gas chromatograph equipped with thermal conductivity detectors (TCD). Hydrogen and helium were analyzed on one channel using a 1/8" x 6' 60/80 mesh Hayesep Q and 1/8" x 3' 60/80 mesh Mol Siev 13x columns in series/bypass configuration with a 1/8" x 18', 45/60 mesh Mol Siev 5A column. Nitrogen carrier gas flow was 15 ml/min and analysis was done isothermally at 35° C in less than 4 minutes. Oxygen + argon, nitrogen, methane and carbon monoxide were analyzed on the other channel using a 1/8" x 6' 80/100 mesh PoroPak N column in series/bypass configuration with 1/8" x 6' 80/100 mesh Mol Siev 13x and 1/8" x 1.3' 80/100 mesh Mol Siev 5A columns. Helium carrier gas flow was 30 ml/min and analysis was done isothermally at 35° C in less than 9.5 minutes.

The liquid nitrogen-condensable CO_2 and C_2 -C_5 fraction was analyzed on a two channel Hewlet Packard 5890 Series II gas chromatograph equipped with a TCD and a flame ionization detector (FID). A 1/8"x 12' 60/80 mesh Hayesep R packed column was used in conjunction with the TCD for analysis of CO_2, N_2, O_2, Ar, and C_1-C_3. Helium carrier gas flow was 30 ml/min. A 50m x .53 mm x 5 μm methyl silicone column was used in conjunction with the FID to analyze C_1-C_{12} hydrocarbons. Helium carrier gas flow was 15

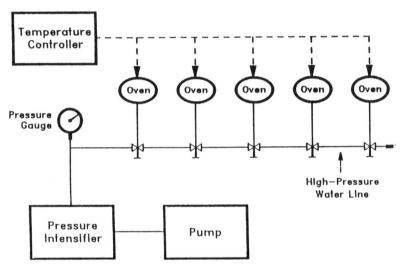

Figure 1: Schematic of high pressure vessel used in confined system pyrolysis.

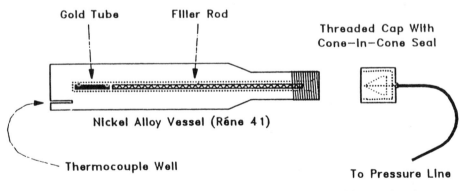

Figure 2: Schematic of hydrothermal laboratory used for confined system pyrolysis.

ml/min. Oven temperature for both channels was programmed from 0° C to 50° C at 10° C/min and 50° C to 200° C at 15° C/min followed by a 5 minute hold. Analysis was complete in 20 minutes.

For the 340° C experiments, gas yields were high enough to permit stable carbon isotope ratio measurements on methane, ethane, propane and carbon dioxide. Individual compounds were isolated and converted to carbon dioxide gas using a Finnigan-MAT semi-automated gas-chromatography-combustion system. In brief, approximately 2 ml of pyrolysis gas was injected onto a 6 ft x 1/8 in stainless steel column packed with 80/100 mesh Poropack N that was temperature programmed from 35° C to 165° C at 10° C/min with a helium flow rate of 30 ml/min. Combustion was achieved in 8 x 3/8 in o.d. sections of Vycor tubing packed with cupric oxide pellets and held at 850° C. Carbon dioxide was separated from water of combustion using standard cryogenic techniques and analyzed for $^{13}C/^{12}C$ ratios with a Finnigan-MAT 251 isotope ratio mass spectrometer. Results are reported in ∂ notation relative to PDB standard and were calibrated assuming $\partial^{13}C(NBS-22) = -29.81$ permil (23).

Liquid Product Analysis. Following gas analysis, the dry ice/acetone bath was removed and the trap rinsed 3 times with pentane to obtain the $C_6 - C_{12}$ gas condensate fraction. This was added to the total bitumen extract. The remaining gold tube was cut open with a scalpel and the residual coal extracted 3 times by sonication in a 4:1 dichloromethane/methanol solvent mixture to obtain the bitumen fraction. Following centrifugation, the extract was decanted and passed through a .45 micrometer nylon filter. Bitumen samples were analyzed on a Hewlett Packard Model 5880 GC equipped with an on-column injector, a DB-1 30 meter, .32 mm i.d., .25 µm methyl silicone capillary column and a flame ionization detector.

Solid Product Analysis. Coal samples were prepared for vitrinite reflectance analysis by first grinding in an agate mortar, mixing the powder with epoxy resin and mounting the mixture on a plexiglas slide in a predrilled 5 mm well. After hardening, the coal/epoxy surface was wet sanded with 600 grit sandpaper and polished successively with 1.0 µm and 0.5 µm alumina.

Vitrinite reflectance (% R_o) was measured at 546 nm wavelength using a Zeiss reflected light microscope. The percent reflected light from fifty telocollinite maceral fragments per sample was measured. The microscope was equipped with a Hewlett Packard 300 computer for statistical manipulation of the data.

Elemental analysis (% C, H, N, S, O) of the residual coals were determined at Huffman Laboratories, Inc. in Golden, Colorado using standard methods.

Results

The gas, liquid and solid product data for the confined pyrolysis of coal at 300° C and 340° C are summarized in Tables I and II, respectively. Since each

Table I: Gas yield, pristane/n-C_{17} ratio and vitrinite reflectance data for coal pyrolysed at 300 °C and various pressures

Sample #	134	135	136	142	143	145	132	133	140	141	147	146	148	149	150
Pressure (bars)	70	70	207	207	345	345	483	483	690	690	1378	1378	2000	2000	2000
N_2 (ml/g)	0.63	2.11	1.17	1.65	1.74	1.32	0.62	0.53	1.63	0.98	1.15	1.32	1.75	0.98	1.63
O_2 + Ar (ml/g)	1.38	0.47	0.06	0.69	0.07	0.04	0.08	0.07	0.93	0.98	0.04	0.06	0.06	0.07	0.05
H_2S (ml/g)	0.38	-	0.15	0.23	0.30	0.08	0.44	0.00	0.19	0.19	0.10	0.31	0.07	0.04	0.05
H_2 (ml/g)	0.43	0.45	0.02	0.02	0.40	0.01	0.01	0.01	0.01	0.01	0.01	0.41	0.01	0.00	0.01
CO (ml/g)	<.01	<.01	<.01	<.01	<.01	<.01	<.01	<.01	<.01	<.01	<.01	<.01	<.01	<.01	<.01
CO_2 (ml/g)	3.77	-	3.27	4.53	3.95	3.41	4.54	4.65	4.65	3.99	3.30	3.13	4.06	2.87	3.92
CH_4 (ml/g)	2.01	1.74	2.11	2.29	2.12	2.13	2.26	2.09	2.16	2.25	1.87	1.44	1.80	1.62	1.60
C2 (ml/g)	0.39	-	0.28	0.42	0.39	0.26	0.34	0.34	0.40	0.32	0.16	0.20	0.16	0.10	0.15
C3 (ml/g)	0.17	-	0.08	0.15	0.21	0.13	0.10	0.10	0.11	0.09	0.04	0.09	0.03	0.02	0.03
iC4 (ml/g)	0.03	-	·0.01	0.02	0.03	0.03	0.02	0.02	0.02	0.01	<0.005	0.02	<0.005	<0.005	<0.005
nC4 (ml/g)	0.04	-	0.01	0.03	0.06	0.03	0.01	0.02	0.01	0.01	<0.005	0.02	<0.005	<0.005	<0.005
iC5 (ml/g)	0.02	-	0.01	0.01	0.03	0.02	0.01	0.01	0.01	0.01	<0.005	0.02	<0.005	<0.005	<0.005
nC5 (ml/g)	0.01	-	<0.005	0.01	0.02	0.01	0.01	0.01	0.01	0.01	<0.005	0.01	<0.005	<0.005	<0.005
C6 (ml/g)	<0.005	-	<0.005	0.01	0.01	<0.005	0.01	0.01	<0.005	<0.005	<0.005	<0.005	<0.005	<0.005	<0.005
Pristane/n-C_{17}	-	1.22	-	1.23	-	-	0.83	-	1.07	-	1.49	-	-	1.64	-
Ro (%)	0.98	0.94	0.99	1.02	0.98	0.97	1.00	0.99	0.98	0.97	0.88	0.85	0.82	0.84	0.84

Table II: Gas yield, pristane/n-C17 ratio and vitrinite reflectance data for coal pyrolysed at 340 °C and various pressures

Sample #	156	159	163	164	167	169	171	173	158	152	153	165	166
Pressure (bars)	70	207	207	345	345	483	483	690	690	1378	1378	2000	2000
N_2 (ml/g)	1.52	1.88	1.44	1.56	1.49	1.45	1.46	1.30	1.06	1.01	1.50	1.35	1.44
$O_2 + Ar$ (ml/g)	0.12	0.67	1.10	1.11	1.71	0.69	0.27	0.12	0.49	0.08	0.11	0.05	1.61
H_2S (ml/g)	1.33	1.15	1.57	1.20	-	0.96	-	0.97	0.86	1.16	0.95	0.87	-
H_2 (ml/g)	0.85	0.91	0.85	0.14	0.12	0.09	-	0.06	0.11	0.03	0.05	0.02	0.02
CO (ml/g)	<0.01	<0.0	<0.01	<0.01	<0.01	<0.01	<0.01	<0.01	<0.01	<0.01	<0.01	<0.01	<0.01
CO_2 (ml/g)	6.81	7.13	7.33	7.88	8.48	7.56	-	7.41	7.31	7.22	6.80	7.22	-
CH_4 (ml/g)	8.13	7.49	8.51	8.22	9.42	9.52	9.04	9.99	9.57	9.75	9.16	8.59	7.87
C2 (ml/g)	1.75	1.66	1.93	2.19	2.14	1.91	-	1.98	1.78	1.92	2.11	1.90	-
C3 (ml/g)	0.76	0.71	0.78	0.82	0.77	0.70	-	0.73	0.64	0.82	0.69	0.61	-
iC4 (ml/g)	0.10	0.11	0.09	0.09	0.09	0.08	-	0.07	0.05	0.04	0.06	0.03	-
nC4 (ml/g)	0.16	0.17	0.16	0.14	0.13	0.13	-	0.14	0.11	0.13	0.16	0.11	-
iC5 (ml/g)	0.07	0.09	0.06	0.04	0.05	0.12	-	0.13	0.11	0.15	0.16	0.12	-
nC5 (ml/g)	0.04	0.05	0.04	0.03	0.03	0.03	-	0.03	0.02	0.03	0.04	0.03	-
C6 (ml/g)	<0.005	0.02	<0.005	0.01	0.04	0.01	-	0.01	0.01	<0.005	<0.005	<0.005	-
Pristane/n-C17	1.41	-	1.49	1.58	-	-	1.26	0.77	-	1.06	-	-	1.08
R_o (%)	1.45	1.44	1.48	1.57	1.55	1.53	1.53	1.50	1.53	1.52	1.56	1.44	1.43

set of experiments was run at constant time and temperature, measured changes in gas yield, pristane/n-C_{17} ratio and vitrinite reflectance reflect changes in the rate of generation or maturation of these components under the influence of pressure. The error bars in the gas yield figures are determined from the reproducibility of various gas standards analyzed by the vacuum line/cryogenic methods described in this paper. The error bars are 6% for methane and hydrogen and 4% for carbon dioxide. Vitrinite reflectance figures represent the pooled standard deviation of the 300° C and 340° C data sets, respectively.

Gas Products.

Methane Yield. At 300° C, methane yield increases from about 1.9 ml/g coal at 70 bars to 2.2 ml/g coal at 690 bars before decreasing to 1.6 ml/g coal at 2000 bars (Figure 3). At 340° C, methane yield increases from approximately 8.0 ml/g coal at 210 bars to 9.7 ml/g coal at 690 bars before decreasing to 8.4 ml/g coal at 2000 bars (Figure 3). This corresponds to an initial increase of 15% to 25% in the rate of methane generation as pressure increases from 70 bars to approximately 600 bars and a decrease of 20% and 25% in the rate of methane generation as pressure increases from approximately 600 bars to 2000 bars. These results demonstrate that the kinetics of methane generation are influenced by pressure.

Hydrogen Yield. Hydrogen gas yield is found to be highest at 70 bars at 300° C and below 210 bars at 340° C. It decreases rapidly as pressure increases above these values (Figure 4). At 300° C, hydrogen yield is 0.44 ml/g coal at 70 bars and decreases to 0.02 ml/g coal at 210 bars and higher pressures. At 340° C, hydrogen yield is roughly constant at 0.85 ml/g coal from 70 bars to 210 bars and decreases with further increase in pressure to 0.01 ml/ g coal at 2000 bars. These data suggest that hydrogen is relatively stable at lower pressures, but reacts rapidly with unsaturated compounds as pressure increases under the confined conditions of these experiments.

Ethane, Propane, Butane and Pentane Yields. Detailed analysis of the C_{2+} hydrocarbon gas data was difficult because of the small sample yields and relatively poor reproducibilities. Nevertheless, differences in the yields of ethane, propane, the butanes and the pentanes are apparent. In the 300° C runs, the yield of each of the measured compounds decreases uniformly from 70 bars to 2000 bars. With increasing pressure, ethane increases in abundance relative to propane (Figure 5) and the other C_{2+} hydrocarbon gases. Gas wetness (ΣC_2-C_6/ΣC_1-C_6), a parameter commonly used in the genetic characterization of natural gases, decreases from .25 at 70 bars to less than .10 at 2000 bars (Figure 5).

Propane, the butanes and n-pentane also decrease with increasing pressure in the 340° C runs. In contrast, however, ethane and isopentane yields increase, the latter by a factor of 3 (.05 ml/g to .15 ml/g; Table II). The contrary trends of ethane and isopentane may indicate the occurrence of bimolecular hydrogen transfer reactions and alkene addition reactions,

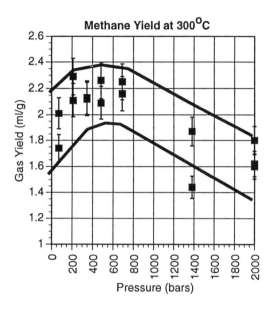

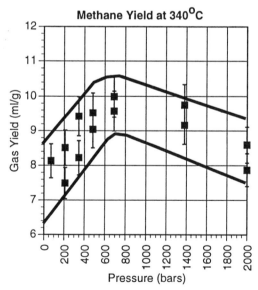

Figure 3: Changes in methane yield with pressure reflecting changes in the rate of generation of methane with pressure.

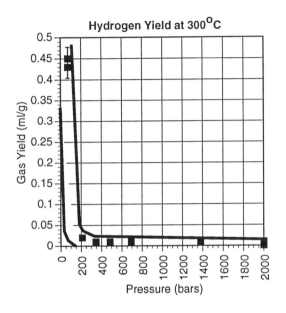

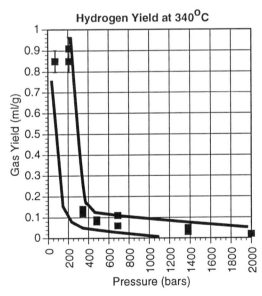

Figure 4: Changes in hydrogen yield with pressure reflecting the consumption of hydrogen through the saturation of alkenes with increasing pressure.

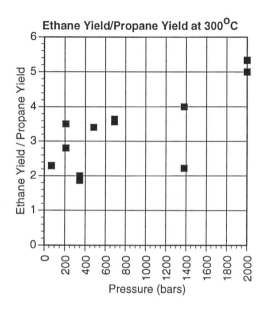

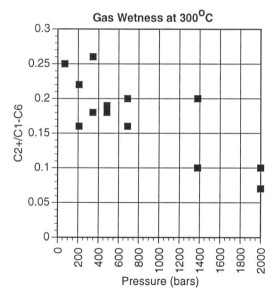

Figure 5: Ethane yield increasing relative to propane with pressure and wetness decreasing as pressure increases at 300° C.

respectively. As in the 300° C runs, ethane increases in abundance relative to propane (Figure 6). Reflecting the different behavior of ethane (and to a smaller extent, isopentane), however, a plot of gas wetness versus pressure is concave upward with a minimum of .23 at 690 bars (Figure 6).

Carbon Dioxide. Carbon dioxide yields increase with increasing pressure to about 600 bars and the decrease with further increase in pressure to 2000 bars for both sets of experiments. At 300° C, carbon dioxide yield increases from roughly 3.8 ml/g at 70 bars to 4.6 ml/g at 480 bars and then decreases to 4.0 ml/g or less at 2000 bars (Figure 7). At 340° C, carbon dioxide yield increases from 6.8 ml/g at 70 bars to 8.2 ml/g at 345 bars then decreases to 7.2 ml/g at 2000 bars. If one assumes that variations in carbon dioxide yield reflect generation rates rather than the influence of secondary CO_2-consuming reactions, then the former must vary by 20% or more as a function of pressure for both sets of experiments.

Carbon Isotope Analysis. Carbon isotope values of methane, ethane, propane and carbon dioxide were determined for selected 340° C experiments and are listed in Table III. Isotopic variability is low: methane $\partial^{13}C$ values average -36.76 ± .09 $^o/_{oo}$, ethane averages -28.15 ± .17 $^o/_{oo}$, propane averages -26.40 ± .33 $^o/_{oo}$ and carbon dioxide averages -19.36 ± .20 $^o/_{oo}$. These data are consistent with data measured for natural gases from coals and dispersed coaly (Type III) material (25-27). As the total carbon isotopic variation observed for each compound is close to analytical precision (approximately $+/_{-}.1$ permil for methane, $+/_{-}.2$ permil for the other gases), it is difficult to discern any influence of pressure. Ethane and propane carbon isotope ratios show no discernible trends. In contrast, methane $\partial^{13}C$ values increase systematically from -36.9 permil at 70 bars to -36.6 permil at 483 bars and appear to decrease again at higher pressures (Figure 8). As a variety of experimental and field data indicate a positive correlation between methane $\partial^{13}C$ values and source rock maturity (28-30), the methane $\partial^{13}C$ data are consistent with other lines of evidence which indicate that pyrolysis pressure is controlling the rate of coal alteration. Although the scatter is too large to draw any conclusions, Figure 8 suggests that a similar pattern may be evident in the carbon dioxide $\partial^{13}C$ data.

Liquid Products.
    Pristane/n-C17 Ratio. Of the various parameters measured from gas chromatograms of the coal extracts (CPI, isoprenoid:n-alkane ratios, various ratios of low and high molecular weight n-alkane, etc.), pristane: n-C_{17} ratio showed the most systematic variation dependence on pressure. Pristane/n-C17 ratios are commonly used as maturation indicators with the pristane/n-C17 ratio decreasing as maturation increases (2). The pristane/n-C17 ratio trends are qualitatively similar at 300° C and 340° C, showing a decrease with increasing pressure to about 600 bars bars and then an increase with further increase in pressure to 2000 bars (Figure 9). This is qualitatively consistent

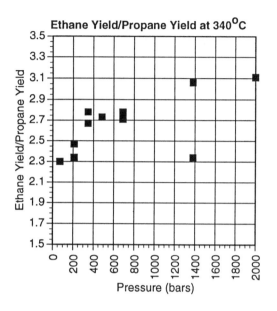

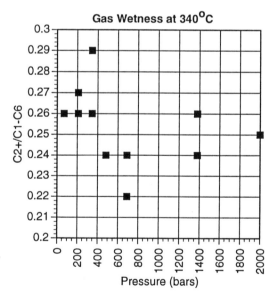

Figure 6: Ethane yield increasing relative to propane as pressure increases and wetness decreasing to about 600 bars and then increasing to 2000 bars at 340° C.

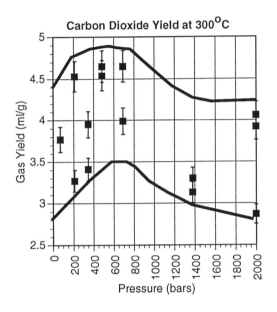

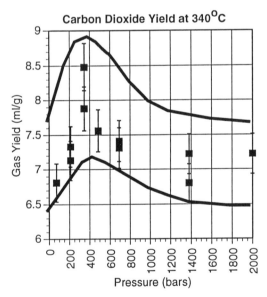

Figure 7: Changes in carbon dioxide yield with pressure reflecting the change in rate of generation with pressure.

Table III: Elemental data for coals pyrolysed at 300° C and 340° C and various pressures

Sample #	134	142	143	141	146	148	156	159	164	169	173	165	IL#6
Temperature (°C)	300	300	300	300	300	300	340	340	340	340	340	340	-
Pressure (bars)	70	207	345	483	1378	2000	70	210	345	483	690	2000	-
Wt % Carbon	62.34	62.70	65.58	63.54	64.86	64.74	65.10	65.34	64.50	64.26	63.30	66.18	77
Wt % Hydrogen	3.92	3.98	4.05	3.90	4.09	4.07	3.49	3.45	3.21	3.22	3.35	3.45	5.7
Wt % Oxygen	8.64	8.64	8.96	8.64	9.92	9.12	6.40	6.40	6.24	6.40	6.08	5.60	10
Wt % Nitrogen	1.26	1.26	1.26	1.26	1.26	1.26	1.40	1.40	1.40	1.40	1.40	1.40	-
H/C ratio	0.75	0.76	0.74	0.74	0.76	0.75	0.64	0.63	0.60	0.60	0.64	0.63	0.89
O/C ratio	0.10	0.10	0.10	0.10	0.11	0.11	0.07	0.07	0.07	0.08	0.07	0.06	0.10

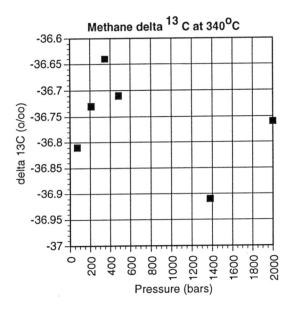

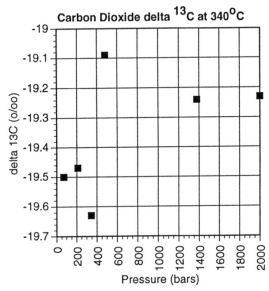

Figure 8: Variations in carbon isotope ratio with pressure for methane and carbon dioxide suggesting pressure subtly influences carbon isotope fractionation.

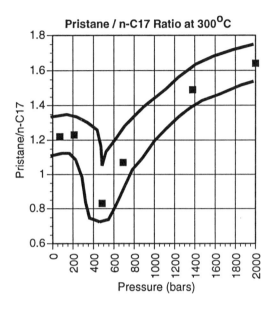

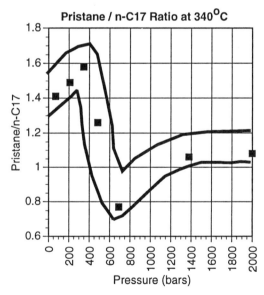

Figure 9: Change in pristane/n-C17 ratio with pressure demonstrating pressure effects this maturation indicator.

with the trends observed for methane and carbon dioxide yields and coal properties (discussed below), namely an apparent increase in maturity with increasing pressure followed by a decrease. The results suggest that pressure influences the rates of pristane and n-C_{17} formation or destruction differently, resulting in the observed trend.

Solid Products.
   Vitrinite Reflectance. At 300° C, vitrinite reflectance increases from %R_o=.96 at 70 bars to %R_o=1.01 at 210 bars and then decreases to %R_o=.84 at 2000 bars (Figure 10). A similar trend is observed for experiments performed at 340° C where vitrinite reflectance increases from %R_o=1.45 at 70 bars to %R_o=1.57 at 345 bars and then decrease to %R_o=1.44 at 2000 bars (Figure 10). The vitrinite reflectance results demonstrate the kinetics of solid transformations as well as liquid and gas product generation are influenced by pressure during sealed gold tube pyrolysis.

   Elemental Analysis. Table IV lists carbon, hydrogen, oxygen and nitrogen analyses performed on the pyrolysed coal residues for the 300° C and 340° C experiments (Table 4). Oxygen to carbon ratios (O/C) were found to be constant with pressure at 300° C and 340° C (Figure 11). The hydrogen to carbon ratio (H/C) shows no change at 300° C, probably due to small amounts of evolved product, and a measurable decrease in the 345 bars to 480 bars pressure range at 340° C. The decrease in H/C ratio presumably results from the increased rates of generation of methane and other hydrocarbon gases as well as bitumen in this pressure range.

   Petrography. Petrographic analysis of product coals reveals physical evidence for the effect of pressure on coal pyrolysis. Figures 12 and 13 contrast the textures of telocollinite for experiments run at 300° C and 340° C at 70 bars, 210 bars, 1380 bars and 2000 bars. The petrography shows vacuoles forming within telocollinite particles for coals pyrolysed at 1380 and 2000 bars, but the lack of such features for coals pyrolysed at pressures below 1380 bars. The vacuoles are similar to those reported by Fowler, et al. (31). They interpreted the features as evidence for the high reactivity of coal during hydrous pyrolysis. The vacuoles within the telocollinite from our experiments suggest pressure is inhibiting the release of pyrolysis products at high pressures. This interpretation is consistent with decreased gas yields, increasing pristane/n-C17 ratio and suppressed vitrinite reflectance observed at high pressures. These results suggest that vacuoles have more than one mode of formation and may indicate different things depending on the type of experiment performed.

Discussion

The coincidence of maxima in methane yield, methane $\partial^{13}C$ and vitrinite reflectance and minima in pristane/n-C17 ratio at approximately 600 bars suggests the same process is governing the generation or maturation of these

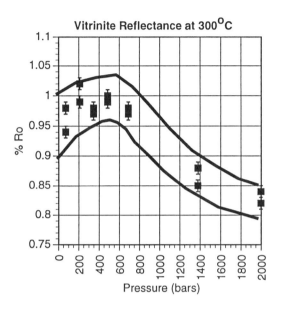

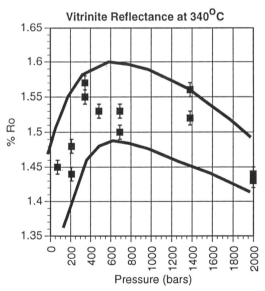

Figure 10: Changes in vitrinite reflectance with pressure reflecting changes in the rate of vitrinite maturation with pressure.

Table IV: Carbon isotope compositions for various gas components relative to PDB Standard*

Sample #	175	176	177	170	180	181	Average	S.D.*
Temperature (° C)	340	340	340	340	340	340		-
Pressure (bars)	70	207	345	483	1378	2000		-
Carbon Dioxide $\partial^{13}C$ (‰)	-19.50	-19.47	-19.63	-19.09	-19.24	-19.23	-19.36	0.20
Methane $\partial^{13}C$ (‰)	-36.81	-36.73	-36.64	-36.71	-36.91	-36.76	-36.76	0.09
Ethane $\partial^{13}C$ (‰)	-28.25	-28.29	-28.22	-28.27	-27.88	-27.98	-28.15	0.17
Propane $\partial^{13}C$ (‰)	-26.48	-26.82	-26.41	-26.62	-26.22	-25.88	-26.40	0.33

*S.D=Standard deviation

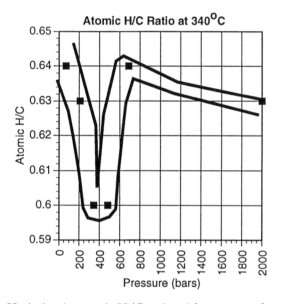

Figure 11: Variation in atomic H/C ratio with pressure showing low H/C corresponding to high methane yield, high vitrinite reflectance and low pristane/n-C_{17}.

Figure 12: Texture of telocollinite from 70 bar (top) and 210 bar (bottom) runs at 300° C.

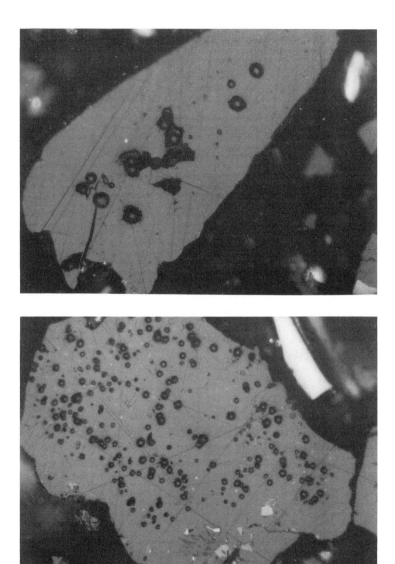

Figure 13: Vacuoles in telocollinite from pyrolysis at 1380 bars and 2000 bars and 300° C reflecting the retardation of product release at high pressures. Similar results are observed at 340° C.

products - maturation of coal. The decrease in atomic H/C in this pressure range coincides with the vitrinite reflectance maximum described above and also correlates with the previously mentioned parameters, confirming earlier suggestions (7) that the kinetic rates of vitrinite maturation, and liquid and gaseous product generation from coal are influenced by pressure.

Illinois Coal #6 contains 8 weight % water (23). Lewan (32) points out that the reactivity of water in hydrous pyrolysis experiments is dependent on the state of the water (vapor < liquid < supercritical fluid). Lewan (32) further noted that the presence of CO_2 can significantly lower the critical point of water, affecting the reactivity. At the temperatures and pressures used in these experiments, the water associated with the coal will be present in the liquid state for all experiments except those run at 340° C and 70 bars where it is probably vapor. To lower the critical point of water to 340° C, the highest temperature used in our experiments, about 18 mol % CO_2 relative to water is required. A maximum of eight mol % CO_2 is generated in our 340° C experiments meaning the water associated with the Illinois Coal #6 never behaves supercritically. Thus, changes in the state of the water associated with the starting coal cannot be invoked to explain the observed results. We therefore attribute the observed changes in gas yield, pristane:n-C_{17} ratio and vitrinite reflectance to the changes in pressure.

Kinetics. The kinetics of coal pyrolysis is a very complicated process. However, an attempt has been made to explain the effect of pressure on the cracking rate of coal using a modification of unimolecular theory. The coal cracking process can be considered a homolytic bond breaking reaction. Consider the general equation

$$A\text{-}B + C \rightleftharpoons (A\cdot + B\cdot)\dagger + C$$
$$(A\cdot + B\cdot)\dagger \rightarrow P$$

where A-B is a coal molecule, C is a colliding molecule, $(A\cdot + B\cdot)\dagger$ is an activated coal species and P is a product. The rate equation describing the formation of product P is derived as follows.

$$d(A\text{-}B)\dagger/dt = [(A\text{-}B)(C)k_1] - k_{-1} (A\text{-}B)\dagger (C) - k_2(A\text{-}B)\dagger$$

If we assume steady state conditions and let $d(A\text{-}B)\dagger/dt = 0$, then
$(A\text{-}B)\dagger = [(A\text{-}B)(C)k_1]/[k_{-1}(C) + k_2]$. It follows that
$d(P)/dt = k_2(A\text{-}B)\dagger = [k_1k_2/(k_{-1}(C) + k_2)] (A\text{-}B)(C) = k_f(A\text{-}B)$
where $[k_1k_2/(k_{-1}(C) + k_2)] (C) = k_f$, the pseudo first order rate constant.

At low pressure, $k_2 >> k_{-1}(C)$ and $k_f = k_1(C)$. Thus, k_f increases as (C) increases with (C) being proportional to pressure. This leads to increasing rates of formation of product P. Our results for experiments performed between 70 bars and 600 bars can be explained by the proposed mechanism and rate equation from the modified unimolecular theory at low pressures. We suggest it is the increased molecular collisions occurring between C (proportional to pressure) and A-B as pressure increases that gives the increased rates of methane and carbon dioxide generation, and increase in vitrinite reflectance observed. Increased A-B + C collisions results in more activated species (A-B)† forming and therefore, more products forming which translates into increased rates of coal cracking.

At high pressure, $k_2 \ll k_{-1}(C)$ and $k_f = (k_1/k_{-1})k_2 = k_{eq}k_2$. Since C does not contribute to k_f, the reaction rate will remain constant at high pressure unless other variables, such as activation volume ($\Delta V\dagger$) are influencing the reaction by influencing k_{eq} or k_2. Results from experiments performed between 600 bars and 2000 bars are consistent with the proposed high pressure reaction mechanism and pseudo first order rate equation from the modified unimolecular theory. We suggest the decrease in rate of generation of methane and vitrinite maturation with pressure reflects the positive activation volume of the coal cracking reaction which becomes significant above 600 bars. This is supported by our estimation of the coal cracking activation volume.

The activation volume for the coal at 300° C and 340° C was estimated to be 7.8 cm^3/mole and 4.6 cm^3/mole, respectively in the 600 bar to 2000 bar pressure range, which are reasonable according to Espenson (33). The calculation was made based on the following assumptions. Relative rate constants for coal maturation (Table V) were calculated from the vitrinite reflectance data assuming the generalized first order reaction A → B describes the process (equation 1).

$$\ln(1 - (B/A_0)) = -k_r t \qquad \text{where } A + B = A_0 = B\infty \qquad (1)$$

In equation 1, A_0 = amount of precursor material available in coal which is consumed when the vitrinite reflectance increases (i.e. to produce an increase in the ordering of the aromatic lamellae). We assume $A_0 = 5 - x$ (the vitrinite reflectance of the starting coal = .2). B = % R_0 (the vitrinite reflectance of the product coal after pyrolysis), k_r = relative rate constant and t = time (seconds). Substituting for A_0 and B in equation 1, we obtain equation 2 which was used to calculate relative rate constants for the confined pyrolysis of coal.

$$\ln(1 - (\% R_0/(5-.2))) = k_r t \qquad (2)$$

Since the $\partial \ln k/\partial P = -\Delta V\dagger/RT$, the activation volumes for the coal cracking reaction at 300° C and 340° C were estimated by plotting $\ln(k_r)$ versus pressure using the 1,380 bars and 2,000 bars rates estimated at each temperature. The $\Delta V\dagger_{300} = 7.8$ cm^3/mole and $\Delta V\dagger_{340} = 4.6$ cm^3/mole. The positive activation volume estimations support the observation that pressure inhibits the maturation process and suggests activation volume becomes the important property influencing the rate of product generation and maturation as pressure increases from about 600 bars to 2000 bars.

C_2-C_5, Hydrocarbons, Carbon Dioxide and Hydrogen Gas. The rate of generation of C2-C5 gases shows no increase at low pressures and is generally

Table V: Relative rate data and calculated activation volume for coals pyrolysed at 300° C and 340° C

Temperature (° C)	300	300	340	340
Pressure (bars)	1380	2000	1380	2000
R_o average (%)	0.875	0.83	1.54	1.435
ln (k_r) (/second)	-14.07	-14.13	-13.41	-13.50
D Va 300 = 4.76 cm^3/mole				
D Va 340 = 7.40 cm^3/mole				

retarded with increase in pressure, except for C_2 and i-C_5 at 340º C. Price and Wenger (*8*) and Blanc and Connan (*14*) show both methane and C2-C5 gas yields decreasing with increasing pressure (i.e., no initial increase in gas yield observed at low pressures). As these workers both used Type II kerogen for their studies, it is conceivable that kerogen type may be influencing the kinetics of product generation.

The pressure dependence of carbon dioxide yield is similar to that observed for methane and can be explained by the same mechanism. Both the 300º C and the 340º C results show a slight increase in yield from 1380 to 2000 bars. However, this may reflect carbon dioxide being generated by decomposition from different functional groups.

The hydrogen yield data show hydrogen gas to be relatively stable at pressures below 70 bars at 300º C and 210 bars at 340º C and to react rapidly as pressure increases above these values. This suggests hydrogenation reactions are favored at higher pressures in the confined system and the lack of olefins detected in gas and liquid pyrolysis products supports this. Domine (*16*) made similar observations from hexane cracking experiments, showing olefin yields decrease as pressure increases during confined pyrolysis.

Liquid Products.   The initial decrease in pristane/n-C_{17} ratio at 300ºC or increase and then decrease in pristane/n-C_{17} at 340ºC with increasing pressure to about 600 bars can be attributed to an increase in the rate of generation of pristane and/or n-C_{17} with pressure or an increase in the rate of destruction of pristane with increasing pressure. As pressure increases from 600 bars (where the pristane/n-C_{17} ratio minimum is observed) to 2000 bars, the pristane/n-C_{17} ratio increases suggesting that activation volume may be inhibiting the generation of n-C_{17} or the destruction of pristane.

Solid Products.   The atomic H/C ratio of pyrolysis residues decreases slightly in the 350-500 bar pressure range, an indication of increased maturation. The H/C data strengthen the argument that pressure is inhibiting the generation or expulsion of gas and bitumen from telocollinite macerals during confined pyrolysis.

Petrography.   The observation under reflected light of bubbles in telocollinite macerals from experiments performed at 300º C and 340º C and pressures of 1380 bars and 2000 bars suggest pressure is inhibiting the release of gas and liquid product from telocollinite during pyrolysis. We believe the bubbles in the telocollinite form from exsolution of gas and liquid products during depressurization of the bombs after the experiments are stopped. The retarded vitrinite reflectance values observed in the high pressure runs are interpreted to result from bitumen retention within the telocollinite. This is consistent with results of other workers who find bitumen retained within vitrinite macerals suppresses vitrinite reflectance (*34-36*).

Application of Results to Sedimentary Basins. In basins with normal geothermal gradients, the onset of oil and gas generation commonly occurs at depths near 3 km. At these depths, fluid pressures may range from roughly 325 bars, corresponding to a hydrostatic gradient, to 600 bars or more, corresponding to a low-permeability shale at 90% of the lithostatic gradient (37-41). In rapidly subsiding basins, such as portions of the U.S. Gulf Coast, the top of the oil window may be considerably deeper and significant gas generating potential may be present in source rocks at depths as great as 6 km (9-13). In geopressured strata, fluid pressures at these depths may approach 1200 bars. The results of our confined coal pyrolysis experiments show that the rates of coal maturation and attendant hydrocarbon generation may vary significantly in this pressure range.

The implications of pressure influencing organic maturation rates in nature are significant. First, our results suggest that in the pressure range of 70 to 600 bars, maturation reaction rates can increase such that hydrocarbon generation may occur at temperatures lower than that normally expected for the onset of oil generation. Whelan (personal communication, 1993) has found vitrinite reflectance to increase in overpressured strata, supporting our laboratory results. Pressure may be a particularly important variable influencing maturation rates in overpressured strata at relatively low temperatures. Second, our results suggest that pressures above 600 bars could retard kerogen maturation and oil generation such that "undermature" situations are observed with respect to burial depth and temperature. This would help account for the preservation of oil in some deep, hot reservoirs (42). Third, basin models for oil and gas generation which do not include pressure as a parameter in kinetic expressions may misrepresent the evolution of vitrinite reflectance and hydrocarbon generation. Gas generation rates appear to vary as much as 20% in the 70 - 600 bars range and suppression of maturation reactions at higher pressures could shift the position of the oil and gas windows.

Conclusions

Results from sealed gold tube pyrolysis experiments on coal shows pressure does influence the rates of gas and liquid product generation and solid maturation, although the effect is secondary. The confined pyrolysis experimental results can be modeled using the modified unimolecular theory. Between 70 bars and about 600 bars, product generation rates and maturation rates increase with increasing pressure. We attribute this rate increase to an increase in molecular collisions resulting in more rapid formation of activated intermediate species which can decompose into gas and liquid products.

Above about 600 bars, pressure retards kerogen maturation and the generation of gas and liquid. We attribute this to the positive activation volume of coal maturation reactions and to pressure inhibition of the decomposition of activated intermediate species.

Our results suggest that maturation rates may be elevated in rapidly deposited, overpressured strata (p<600 bars) such that hydrocarbon generation occurs at temperatures lower than normally expected. In deep overpressured, high-temperature rocks (P>>600 bars), however, rates of maturation may be retarded and oil preserved to depths greater than normally expected. Basin models for hydrocarbon generation and destruction could be improved by incorporating reaction rate equations which reflect the influence of fluid pressure.

Finally, the results of our experiments show pressure slightly influences carbon isotopic fractionation during the pyrolysis of coal.

Acknowledgments

We would like to thank Chevron Petroleum Technology Company management for its support of this research effort, Mark Haught for his extensive help in setting up and maintaining the GC-vacuum line system, Tan Ta for assistance in vitrinite reflectance analysis and Greg Hobson for assisting with GC gas analysis. We would like to thank the department of Earth and Space Sciences at UCLA for access to the hydrothermal lab. Thanks also to Martin Fowler and Patrick Hatcher for helpful comments and criticisms.

Literature Cited

1. Hunt, J. (1979) *Petroleum Geochemistry and Geology*, 617 pp. Freeman, San Fransisco.
2. Tissot, B.P., and Welte, D.H. (1984) *Petroleum Formation and Occurrence*. 699 pp. Springer-Verlag, New York.
3. Hesp, W. and Rigby, D. (1973) The geochemical alteration of hydrocarbons in the presence of water. *Erdol und Kohle-Erdgas 26* 70-76.
4. McTavish, R.A. (1978) Pressure retardation of vitrinite diagenesis, offshore northwest Europe. *Nature 271* 648-650.
5. Cecil, B., Stanton, R., Allshouse, S., and Cohen, A. (1979) Effects of pressure on coalification. *Ninth International Congress of Carboniferous Stratigraphy and Geology*, University of Illinois, Urbana, Illinois, May 19-26, Abstracts of Papers, p. 32.
6. Horvarth, Z.A. (1980) Study on the maturation process of huminitic organic matter by means of high-pressure experiments. *Acta Geol. Hung. 26* 137-148.
7. Sajgo, C., McEvoy, J., Wolff, G.A. and Horvarth, Z.A. (1986) Influence of temperature and pressure on maturation processes - I. Preliminary report. *Advances in Organic Geochemistry 10* 331-337.
8. Price, L.C. and Wenger, L.M. (1992) The influence of pressure on petroleum generation and maturation as suggested by aqueous pyrolysis. Advances in *Org. Geoch. 19* 141-160
9. Price, L.C., Clayton, J.L., and Rumen, L.L. (1979) Organic geochemistry of a 6.9 km deep well, Hinds County, Mississippi. *Gulf Coast Assoc. and Geol. Soc. Trans. 29* 352-370.

10. Price, L.C., Clayton, J.L., and Rumen, L.L. (1981) Organic geochemistry of the 9.6 km Bertha Rogers #1, Oklahoma. *Org. Geoch. 3* 59-77.

11. Price, L.C. (1981) Organic geochemistry of 300o C, 7 km core samples, South Texas. *Chem. Geol. 37* 205-214.

12. Price, L.C. (1988) The organic geochemistry (and causes thereof) of high-rank rocks from the Ralph Lowe-1 and other well bores. *USGS Open-File Report 91-307* 55p.

13. Price, L.C., and Clayton, J.L. (1990) Reasons for and significance of deep, high-rank hydrocarbon generation in the south Texas Gulf Coast. In *Gulf Coast Oils and Gases:* Gulf Coast Section, Society of Economic Paleontologists and Mineralogists, Ninth Annual Research Conference Symposium Volume. (Edited by D. Schumacher and B.F. Perkins) pp. 105-138.

14. Blanc, P. and Connan, J. (1992) Generation and expulsion of hydrocarbons from a Paris Basin Toarcian source-rock: An experimental study by confined-system pyrolysis. *Energy and Fuels 6* 666-677.

15. Perlovsky, L.I. and Vinkovetsky, Y.A. (1989) The role of pressure in the generation and preservation of hydrocarbons. *Boll. di Geof. Teor. Appl. 122* 87-90.

16. Domine, F. (1989) Kinetics of hexane pyrolysis at very high pressures. 1. Experimental study. *Energy and Fuels 3* 89-96.

17. Monthioux, M., Landais, P. and J.C. Monin (1985) Comparison between natural and artificial maturation series of humic coals from the Mahakam delta, Indonesia. *Org. Geoch. 8* 275-292.

18. Monthioux, M., Landais, P. and Durand, B. (1986) Comparison between extracts from natural and artificial series of Mahakam delta coals. In Advances in Organic Geochemistry 1985 (Edited by D. Leythaueser and J. Rullkotter). *Org. Geoch. 10* 299-311.

19. Braun, R.L. and Burnham, A.K. (1990) Mathematical model of oil generation, degradation, and expulsion. Energy and Fuels 4 132-146.

20. Doue, F. and Guiochon, G.J. (1968) *J. Chim. Phys. 65* 395-409.

21. Fabuss, B.M., Smith, J.O. and Satterfield, C.N. (1964) In *Advances in Petroleum Chemistry and Refining,* ed. McKetta, J.J. 157-201.

22. Shin, S.-C., Baldwin, R.M. and Miller, R.L. (1989) Correlation of bituminous coal hydroliquification activation energy with fundamental coal chemical properties. *Energy and Fuels 3* 193-199.

23. Burnham, A.K., Oh, M.S. and Crawford, R.W. (1989) Pyrolysis of Argonne Premium Coals: activation energy distributions and related chemistry. *Energy and Fuels 3* 42-55.

24. Lewan, M.D. (1985) Evaluation of petroleum generation by hydrous pyrolysis experimentation. *Phil. Trans.. R. Soc. Lon. 315* 124-134.

25. Jenden, P.D. and Kaplan, I.R. (1989) Origin of natural gas in Sacramento Basin, California. *Am.. Assoc. Pet. Geol. 73* 431-453.

26. Jenden, P.D. and Kaplan, I.R. (1986) Comparison of microbial gases from the Middle America Trench and Scripps Submarine Canyon: Implications for the origin of natural gas. *Appl. Geoch. 1* 631-646.

27. Schoell, M., Faber, E., and Coleman, M.L. **(1983)** Carbon and hydrogen isotopic compositions of the NBS 22 and NBS 21 stable isotope reference materials: an inter-laboratory comparison. *Org. Geoch. 5* 3-6.
28. Schoell, M. **(1983)** Genetic characterization of natural gases. *Am. Assoc. Pet. Geol. 67* 2225-2238.
29. Chung, H.M. and Sackett, W.M. **(1979)** Use of stable carbon isotope compositions of pyrolytically derived methane as maturation indices for carbonaceous material. *Geochim. Cosm. A cta 43* 1979-1988.
30. Faber, E. **(1987)** Zur Isotopengeochemie gasforminger Kohlen-wasserstoffe. *Erdol Erdgas Kohle 103* 210-218.
31. Fowler, M.G., Gentzis, T., Goodarzi, F., and Foscolos, A.E. **(1991)** The petroleum potential of some Tertiary lignites from Northern Greece as determined using hydrous pyrolysis and organic petrologic techniques. *Org. Geoch.17* 805-826.
32. Lewan, M.D. **(1993)** Laboratory simulation of petroleum formation: hydrous Pyrolysis. In *Organic Geochemistry* 1993 (Edited by M.H. Engel and S.A. Mack) 419-442.
33. Espenson, J.H. **(1981)** *Chemical Kinetics and Reaction Mechanisms.* McGraw-Hill, New York 218 p.
34. Suarez-Ruiz, I., Iglesias, M.J., Bautista, A.J., Laggoun-Defarge, F., and Pardo, J.G. **(1993)** Petrographic and geochemical anomalies detected in the Spanish Jurassic jet. *206th ACS National Meeting Book of Abstracts*
35. Gentzis, T., Goodarzi, F. and Snowdon, L.R. **(1993)** Why is reflectance "suppressed" in some Cretaceous coals from Alberta. *206th ACS National Meeting Book of Abstracts*
36. Quick, J.C. **(1993)** Isorank variation of vitrinite reflectance and fluorescence intensity. *206th ACS National Meeting Book of Abstracts*
37. Powers, M.C. **(1967)** Fluid-release mechanisms in compacting marine mudrocks and their importance in oil exploration. *Am. Assoc. Pet. Geol. 51* 1240-1254.
38. Burst, J.F. **(1969)** Diagenesis of Gulf Coast clayey sediments and its possible relation to petroleum migration. *Am. Assoc. Pet. Geol. 53* 73-93.
39. Perry, E. and Hower, J. **(1972)** Late-stage dehydration in deeply buried pelitic sediments. *Am. Assoc. Pet. G eol. 56* 2013-2021.
40. Bruce, C.H. **(1984)** Smectite Dehydration - Its relation to structural development and hydrocarbon accumulation in Northern Gulf of Mexico Basin. *Am. Assoc. Pet. G eol. 68* 673-683.
41. Freed, R.L. and Peacor, D.R. **(1989)** Geopressured shale and sealing effect of smectite to illite transition. *Am. Assoc. Pet. G eol. 73* 1223-1232.
42. Price, L.C. **(1982)** Organic geochemistry of core samples from an untra-deep hot well (300° C, 7 km). *Chem. Geol. 37* 215-228.

RECEIVED April 15, 1994

Chapter 12

Evolution of Vitrinite Ultrafine Structures During Artificial Thermal Maturation

F. Laggoun-Défarge[1], E. Lallier-Vergès[1], I. Suárez-Ruiz[2], N. Cohaut[3],
A. Jiménez Bautista[2], P. Landais[4], and J. G. Prado[2]

[1]Unité de Recherche en Pétrologie Organique, Unité de Recherche
Associée, 724 du Centre National de la Recherche Scientifique,
Université d'Orléans, 45067 Orléans, Cedex 2, France
[2]Instituto Nacional del Carbón (CSIC), La Corredoria, s/n. Apartado 73,
33080 Oviedo, Spain
[3]Centre de Recherche sur la Matière Divisée, Unité de Marine
Recherche, 0131 du Centre National de la Recherche Scientifique,
Université d'Orléans, 45067 Orléans, Cedex 2, France
[4]Centre de Recherche sur la Géologie de l'Uranium, BP 23,
54501 Vandœuvre-lès-Nancy Cedex, France

Simulation of the natural coalification of a low-rank pure vitrinite
(Rm=0.5%) was performed in a confined-pyrolysis system. Vitrinite-
loaded gold tubes were isothermally heated at different temperatures
(300 to 450ºC) during 72 hours at a constant pressure (70 MPa). At the
end of each pyrolysis run, vitrinite samples were analyzed using
reflectometry, Rock-Eval pyrolysis, X-ray diffraction and principally
transmission electron microscopy (TEM).
TEM investigations of ultra-thin sections show that thermal degradation
initially affects the homogeneous vitrinite relative to cell walls which are
more resistant. This degradation starts at 320ºC with fracturing and the
appearance of uranium aggregates systematically associated with
degraded vitrinite. The peak of the plastic phase is reached at 340ºC,
corresponding to maximum hydrocarbon generation (extract yield =
9.3%). Botanical structures identified in the initial sample disappear at
380ºC. No organisation of the basic structural units has been noticed
even at 450ºC.

Vitrinites are the major petrographic components of most coals, and thus contribute
significantly to their industrial properties. Coal petrography classically describes some
major types of vitrinites, essentially based on their morphological properties observed
in light microscopy. Due to the limited resolution of these optical methods, the ultrafine
structures of vitrinites (porosity, mineral inclusions, vegetal structures...) are not yet
well defined. However, some recent studies using electron microscopy have been
conducted on the main types of macerals in order to define their composition and
textural aspect (1-6). This study addresses this problem and has two main aims. The
first is to investigate the ultrafine structures of a low-rank, pure vitrinite. This vitrinite
was hand-picked in large quantities from the *same* layer, which allowed a suite of
different analyses to be performed. The second is to characterize the behaviour of
vitrinite, and more precisely to follow the evolution of its fine and ultrafine structures
during thermal maturation. Due to the difficulty of sampling a complete and

homogeneous natural maturation series of coals, an experimental simulation of natural coalification was used (confined pyrolysis) which has been shown to reproduce natural maturation (*7, 8*).

Sampling, Preparation Methods and Analytical Techniques

A vitrinite sample was selected by hand-picking from a Stephanian high-volatile bituminous coal. This sample comes from the uppermost seam of the open pit mine named Fouthiaux in the Montceau-les-Mines Basin in France. The choice of this vitrinite was based on its homogeneity in light microscopy, its low mineral content and its relative immaturity.

Artificial maturation of this vitrinite was carried out in a confined pyrolysis system. 300 mg of a powdered sample were placed under argon atmosphere inside a sealed gold tube which was introduced into a cold-seal autoclave. The standard conditions of heat treatment used were, a 72 hour isothermal stage, temperatures of 300, 320, 340, 380, 400 and 450ºC, and a constant pressure of 70 MPa which was exerted on the tube in order to ensure a confinement (*9, 10*).

Both the raw vitrinite and pyrolized samples were studied on polished sections and/or embedded grain sections by light microscopy (reflected light and under UV excitation). Additional aliquots were ground in order to be studied by Rock-Eval pyrolysis, X-ray diffraction, scanning electron microscopy (SEM) and finally, transmission electron microscopy (TEM).

TEM observations were performed on ultra-thin sections using a STEM Philips CM-20 instrument. Elemental microanalyses were obtained using an energy dispersive X-ray (EDX) spectrometer fitted to the STEM CM-20. The ultrathin sections (30 to 50 nm thick), which were cut using a diamond knife, from grains embedded in a methyl-metacrylate resin (*11*), were placed on perforated carbon filmed grids. The pyramid plane surfaces, from which ultrathin sections had been made, were observed by optical microscopy and by SEM, in order to compare these observations with those of the TEM.

The X-Ray Diffraction (XRD) apparatus consists of a transmission diffractometer using $K_1 \alpha$ molybdenum radiation, equipped with a curve localization detector connected to a multichannel analyser. Relative intensities are expressed in arbitrary units. It is from the (002) reflection, which is due to the periodic stacking of aromatic layers, that the following parameters are calculated:
(1) d_{002} which represents the mean interlayer spacing of aromatic sheets:

$$2 \, d_{002} \sin \theta = \lambda$$
$\lambda = 0.709 \, \text{Å}$ (wave length for Molybdenum radiation)

(2) L_c which is the stack height of aromatic layers deduced from the Scherrer formula:

$$L_c = 0.89 \, / \cos \theta \, \Delta \, (2 \, \theta)$$

(3) <N> which is the averaged number of aromatic layers:

$$<N> = L_c \, / \, d_{002}$$

The (10) band, which corresponds to the extension of aromatic molecules in the plane of the layer, has also been studied.

Results and Discussion

Petrographical and Geochemical characteristics. The raw vitrinite is composed of 97 vol. % telocollinite and humotelinite and is relatively immature (Rm =

0.5% and Rock-Eval pyrolysis Tmax = 438ºC). The hydrogen index (HI), about 104 mg HC/g C org. (Table I), is typical of type III organic matter.
Both reflectance and Rock-Eval pyrolysis Tmax increase as confined pyrolysis temperature (P. T.) increases during maturation, up to 2.7% for Rm and 590ºC for Tmax at 450ºC P. T. (Table I). Conversely, the HI decreases to 9 mg HC/g C org. at this pyrolysis temperature.

Table I. Reflectance and Geochemical Parameters of the Raw and Pyrolyzed Samples

Samples	R_m(%)	Tmax (ºC)	HI (mg HC/g Corg)	CHCl₃ extract (%)
raw	0.53	438	104	--
300ºC	0.95	467	87	1.76
320ºC	1.18	479	83	5.08
340ºC	1.51	495	69	9.30
380ºC	2.02	549	57	1.07
400ºC	2.19	568	24	0.90
450ºC	2.71	590	8	0

Unlike natural series, the artificially maturated vitrinites studied do not exhibit any anisotropy in polarized light. This is probably due to the multidirectional pressure exerted during the confined pyrolysis.

Evolution of Physical Structure. The evolution of physical structure of the carbonaceous matter has been followed during the artificial maturation by XRD study. The diffractograms of the raw vitrinite, as well as those of the 340 and 450ºC stages, are presented in figure 1 and the structural parameters, calculated from the (002) reflection, are summarized in table II.
The XRD results do not reveal any mineral phases in the vitrinite. The carbonaceous matter of the raw sample is poorly organized with a weak (002) reflection and no (10) band (Figure 1).
The critical stages in the organization of the carbonaceous matter are 340 and 450ºC. At 340ºC, the (002) reflection is more prominent (Figure 1). An even higher degree of organization of turbostratic layers is observed at the 450ºC stage, as the (002) reflection becomes sharper and shifts to higher angles. In addition, the (10) band becomes more evident at 450ºC, which means that the size of basic structural units is increasing (Figure 1).
The mean interlayer spacing of aromatic sheets (d_{002}) decreases from 3.65 Å in the raw sample to 3.54 Å at the 450ºC stage (Table II). Conversely, the mean stacking size of aromatic layers (L_c) and their number (<N>) increase respectively from 11.5 Å and 3 in the raw sample, to 28 Å and 8 at the 450ºC stage (Table II).

Table II. Structural Parameters of the Carbonaceous Matter Calculated from X-ray Diffraction Data

Samples	d_{002} (Å)	L_c (Å)	<N>
raw	3.65	11.5	3
340ºC	3.58	21.4	6
380ºC	3.55	24.5	7
450ºC	3.54	28	8

These results are interpreted as a separation of volatile aliphatic hydrocarbons at 340ºC. This is confirmed by the chloroform extraction results which show that

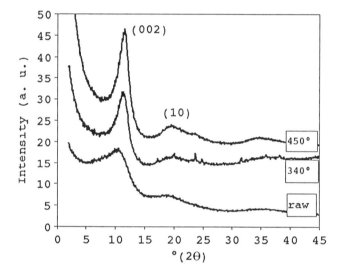

Figure 1: X-ray diffractograms of the raw vitrinite and those pyrolized at 340 and 450ºC.

maximum hydrocarbon generation (extract yield = 9.3%) is acheived at this temperature. An even greater aromatization of the carbonaceous matter is attained at 450ºC.

Evolution of Ultrafine Structure. While the observations in light microscopy have shown that the raw vitrinite is homogeneous and amorphous (Figure 2), the TEM investigations reveal several structures which take the form of sinuous, continuous laminae, denser (more opaque to electrons) than the surrounding material (Figure 3). They have been considered as indicating the presence of vegetal cell walls (5). The latter delimit the cellular cavities which are filled with an amorphous matrix of variable porosity (*see* A and B in Figure 3). Therefore, the differences observed between cell walls and cavities underline the variability of the botanical origin, while the textural variations between the matrices might illustrate differences in the early gelification processes.

The evolution of the fine and ultrafine structure of this vitrinite during artificial coalification was studied by both SEM and TEM.

At 300ºC, the change in the structure is not substantial. However, at 320ºC, the SEM observations show that the sample is more fractured with a porosity ranging from 1 to 10 μm. Several "melted-forms" appear, probably indicating primary hydrocarbon generation. TEM investigations reveal that vegetal structures are still present (*see* cw in Figure 4). However, the amorphous matrices described above (*see* m in Figure 4) are more porous and consist of a granular matter exhibiting polygonal structures (Figure 5). This degraded material typically has pore sizes ranging from a few Ångstroms to 0.2 μm. Therefore, this granular ultrastructure associated with the fracturing is probably the result of thermal degradation, which mostly affects the massive matrix, rather than the more resistant cell walls.

Opaque grains measuring between 10 and 50 Å (Figure 6) are systematically associated with the polygonal structures described above. These grains were identified by EDX analysis as uranium inclusions (*see* inset in Figure 6). These inclusions, being probably in the form of organo-uranyl complexes (*12*), were not detected in the raw vitrinite. Due to the effect of temperature, the transformation of these complexes lead to a breakage of organo-metallic cross-linkages, followed by a formation of uraninite crystals (*13*). Indeed, it has been demonstrated that alcoholic and aliphatic hydrocarbon groups in the organic matter are generally responsible for this reduction (*14*). This reduced uranium may be thus expelled and preferentially concentrated in the zones of vitrinite degradation; i.e., in the granular phase described above.

At 340ºC, devolatilization vacuoles and matrix "melted-forms" appear in large quantity (Figure 7). They certainly indicate the maximum of the plastic phase and thus the peak of hydrocarbon generation, which is confirmed by the chloroform extraction results as discussed above.

At 380ºC, the botanical structures observed below this temperature disappear. The uraninite crystals are well-shaped, often aggregated (*see* ua in Figure 8) and systematically associated with the vacuoles. At this temperature, these crystals are larger in size (200 to 500 Å). A selected-area electron diffraction (SAD) pattern was obtained (*see* inset in Figure 8) and enabled the identification of uraninite crystals.

At 400ºC, no substantial evolution of the ultrafine structure is noticed. However, At 450ºC, SEM observations show that the devolatilization vacuoles have spread and coalesced (Figure 9). Most of these vacuoles were sealed by thin skins which often appear broken, probably due to sudden gas expulsion. This gas production-expulsion phenomenon seems to occur in successive stages as seen in Figure 10. Furthermore, TEM investigations using point high resolution show that the ultrafine structure of vitrinite (*see* v in Figure 11) is still "amorphous" and does not reveal any organisation of basic structural units, even near the rare minerals (clays) detected at this temperature (*see* c in Figure 11), despite their demonstrated catalytic

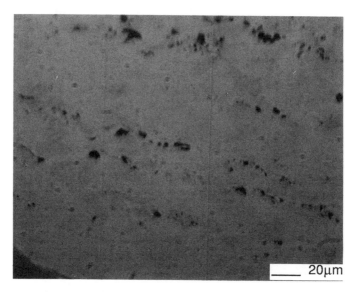

Figure 2: The raw vitrinite observed by reflected light microscopy, oil immersion.

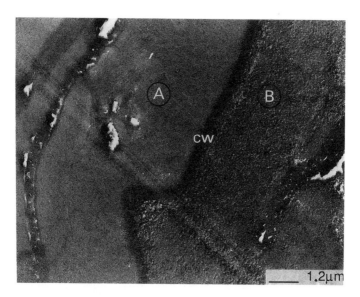

Figure 3: TEM micrograph of the raw vitrinite showing vegetal cell walls (cw) which delimit cellular cavities filled with amorphous matrices of variable porosity (A and B).

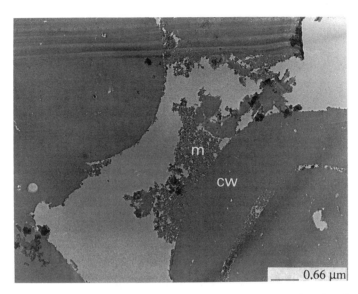

Figure 4: TEM micrograph of the vitrinite pyrolized at 320°C showing that the thermal degradation affects essentially the massive matrices (m) compared to the more resistant cell walls (cw).

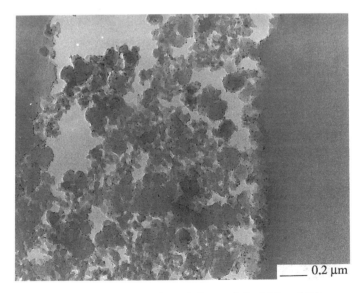

Figure 5: Detail of the thermally degraded material which exhibits a granular structure intimately mixed with more opaque uranium granules.

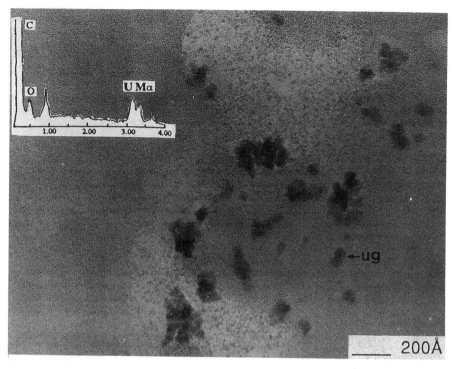

Figure 6: High resolution TEM micrograph of the opaque uranium granules (ug) identified in the X-ray spectrum (see inset).

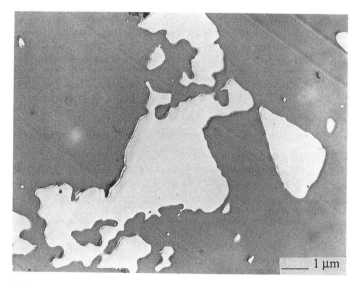

Figure 7: TEM micrograph of devolatilization vacuoles of the vitrinite pyrolyzed at 340ºC.

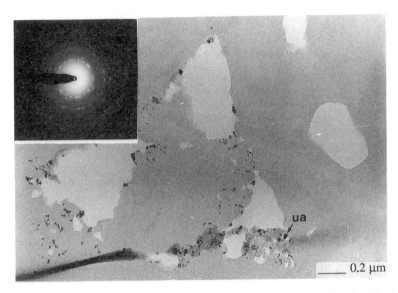

Figure 8: TEM micrograph of uranium aggregates (ua) associated with the devolatilization vacuoles in the vitrinite pyrolyzed at 380ºC, and selected-area electron diffraction (SAD) pattern of these aggregates (see inset).

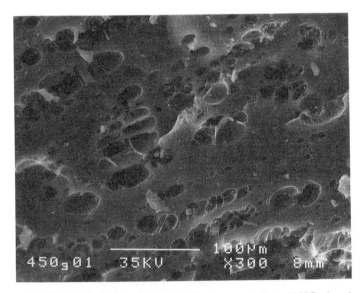

Figure 9: SEM micrograph of the vitrinite maturated at 450ºC showing the generalization of coalescent devolatilization vacuoles.

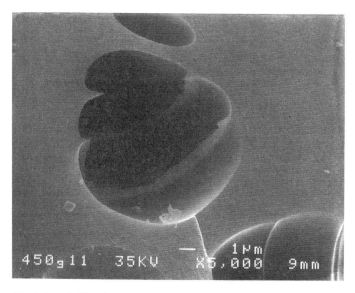

Figure 10: A devolatilization vacuole with two thin broken skins indicating, (1): a sudden gas expulsion, (2): two successive stages of this gas production-expulsion phenomenon.

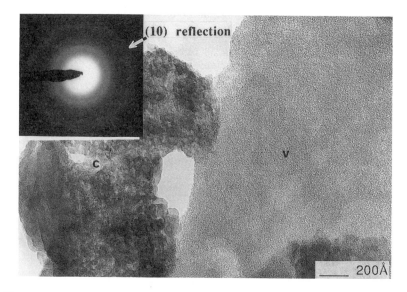

Figure 11: High resolution TEM micrograph of the carbonaceous matter of the vitrinite (v) pyrolyzed at 450°C and its selected-area electron diffraction (SAD) pattern (see inset). This vitrinite is associated with clays (c).

effect. This is indeed supported by the carbonaceous matter SAD pattern which exhibits a very weak (10) reflection (*see* inset in Figure 11).

Conclusions

Electron microscope investigations complemented principally by X-ray diffraction provided furthur information about the ultrafine structure of vitrinites, thus enabling a better understanding of their behaviour during artificial thermal maturation.

Raw vitrinite, which appears completely homogeneous under the light microscope, has been shown by TEM observations, to possess botanical structures. These structures take the form of cell walls delimiting cavities full of amorphous matrices which present different porosities. This was interpreted as a selective preservation of the cell walls, and as differences in the early gelification processes concerning the matrices.

During artificial coalification, it has been shown that thermal degradation begins at 320°C and mostly affects the massive matrices compared to the more resistant cell walls. This degraded vitrinite exhibits a granulous texture and is systematically associated with uranium granules ranging from 10 to 50 Å in size. These inclusions, which are not detected in the raw vitrinite, probably result from breakage of organo-metallic cross-linkages and a formation of uraninite crystals, which are then preferentially concentrated in the zones of vitrinite degradation.

This study shows that the maximum of hydrocarbon generation is reached at 340°C, as indicated by the chloroform extraction results and confirmed by the apppearance of devolatilization vacuoles in TEM. The botanical structures disappear completly at 380°C, replaced by an extensive development of vacuoles which are associated with uraninite aggregates.

Finally, even at 450°C the ultrafine texture of vitrinite does not exhibit any molecular orientation domains at high resolution, and remains as poorly organized elementary carbonaceous granulations. However, the X-ray diffraction study shows that the degree of organization of the carbonaceous matter in turbostratic layers increases (L_c = 28 Å against 11.5 Å in the raw vitrinite). This is probably due to the fact that the lateral organization of the basic structural units is not sufficiently developped to be detected by TEM-high resolution.

Acknowledgments

We are indebted to the European Community, program n° 7220-EC-757 for financial support. We thank A. Genty from the "Ecole Supérieure de l'Energie et des Matériaux", Orléans, and C. Clinard from the "Centre de Nanoscopie Analytique", Orléans, for their technical assistance, and also acknowledge S. Bonnamy and F. Lambert from the "Centre de Recherche sur la Matière Divisée", Orléans, for the ultramicrotom facilities. M. Boussafir, G. Drouet, E. Jolivet and A. Patience are gratefully acknowledged for their assistance in the preparation of the manuscript..

Literature Cited

1. Harris, L.A.; Yust, C.S. *In* : Coal structure, *Am. Chem. Soc.* **1981**, pp.321-336.
2. Lallier-Vergès, E.; Bertrand, P.; Guet, J.M.; Clinard, C.; Lin, Q.; Wu, X.Q. *Bull. Soc. géol. France* **1991**, *162*, pp.163-174.
3. McCartney, C.T.; O'Donnell, H.J.; Ergun, S. *"Coal Science"*, Adv. Chem. ser. **1966**, *55*, pp.261-273.
4. Taylor, G.H. *"Coal Science"*, Adv. Chem. ser. **1966**, *55* pp.274-283.
5. Taylor, G.H.; Shibaoka, M.; Liu, S. *Fuel* **1982**, *61*, pp.1197-1200.
6. Taylor, G.H.; Teichmuller, M. *Int. J. Coal Geol.* **1993**, *22*, pp.61-82.
7. Monthioux, M.; Landais, P.; Monin, J.-C. *Org. Geochem.* **1985**, *8*, pp.275-492.

8. Landais, P.; Zaugg, P.; Monin, J.-C.; Kister J.; Muller, J.-F. *Bull. Soc. géol. Fr.* **1991**, *162*, pp.211-217.
9. Landais, P.; Michels, R.; Poty, B. *J. Anal . Appl. Pyr.* **1989**, *16*, pp.103-115.
10. Landais, P.; Michels, R.; Kister, J.; Dereppe, J.-M.; Benkhedda, Z. *Energy & Fuels* **1991**, *5*, pp.860-866
11. Tchoubar, C.; Rautureau, M.; Clinard, C.; Ragot, J.-P. *J. Microscopie* **1973**, *6*, pp.147-154.
12. Disnar, J.-R.; Trichet, J. *C. R. Acad. Sc. Paris.* **1983**, *296*, pp.631-634.
13. Meunier, J.-D.; Landais, P.; Pagel, M. *Geochem. Cosmochim. Acta* **1990**, *54*, pp.809-817.
14. Nakashima, S.; Disnar, J.-R.; Perruchot, A.; Trichet, J. *Bull. Minéral.* **1987**, *110*, pp.227-234.

RECEIVED April 15, 1994

Chapter 13

Pyrolysis–Carbon-Isotope Method

Alternative to Vitrinite Reflectance as Kerogen Maturity Indicator

William M. Sackett, Zhenxi Li, and John S. Compton

Department of Marine Science, University of South Florida,
St. Petersburg, FL 33701

The pyrolysis-carbon isotope method for determining the thermal maturity of kerogen is based on the exhaustive pyrolysis of whole rock samples and measurement of the amounts and carbon isotope compositions of the pyrolysis- derived methane and the total organic carbon. Potential problems were 1) the pyrolysis time which has been reduced here to one hour at 700°C 2) possible exchange between methane and carbon dioxide, shown here to be unimportant and 3) the stability of methane, which is also shown to be immaterial for the standard time and temperature. Results for a suite of core samples from the Monterey Formation of the Point Arguello field offshore southern California are consistent with other maturity parameters.

The pyrolysis-carbon isotope (PCI) method for determining the maturity of kerogen is based on the exhaustive pyrolysis of whole rock samples and measurement of the amounts and stable carbon isotope compositions of the pyrolysis-derived methane and total organic carbon in the rock sample. To insure that oil staining or migrated oil does not interfere it is best to pre-extract the ground sample with a solvent such as dichloromethane. An amount of ground sample containing about 5 mgs. carbon is placed in a 9 mm o.d. Vycor tube with one end previously sealed and followed by a plug of quartz glass wool to prevent loss of a powdered sample during evacuation. The sample is preheated at 400°C for 10 minutes while evacuating to drive off most of the water in the sample. After cooling, the glass tube is sealed while under vacuum with a natural gas-oxygen torch. The sample tube is then placed in oven at 700°C for one hour. Timing is not started until the sample is at 700°C. Initially, a whole suite of gases is generated at lower temperatures but the only gaseous hydrocarbon that can be found after this treatment is methane. The procedures for combustion and manometric determination of the carbon dioxide from the methane nd kerogen are given in (1) and (2), respectively.

0097–6156/94/0570–0206$08.00/0

The model for interpreting the data was presented in (3) and is reproduced in (Figure 1). Basically it shows that after organic matter is deposited in a basin and is buried below the zone of microbial activity where only thermal processes become important, carbon-12 rich methane is generated as carbon-12-carbon-12 bonds are broken preferentially to carbon-12-carbon-13 bonds. A review of carbon isotope effects associated with the thermogenic formation of methane is given in (4). Referring to (Figure 1), natural processes move the CH_4/C mole ratio from the initial to the present-day position and pyrolysis takes it from there to the postpyrolysis position. The same holds for $\Delta^{13}C$. When plotted one against the other, the closer the today position is to the initial position, the more immature is the kerogen and vice versa. Actually CH_4/C and $\Delta^{13}C$ are independent parameters. One or the other or both may be used as a kerogen maturity indicator. Our past experience suggests that the CH_4/C mole ratio may be somewhat better. As this measurement does not require an isotope ratio mass spectrometer but only a vacuum line, Toepler pump, combustion tube and furnace, calibrated manometer, another temperature controlled furnace for the pyrolyses and ancillary equipment, workers may prefer this simplified procedure.

Because of safety considerations pyrolysis tubes are shielded in stainless steel pipes. The pressure inside the glass tubes reaches a maximum of about 4 atmospheres at 700°C. Very few experiments have blown up. A rare problem is leakage of the seal which may be easily detected by the absence of black pyrolysis products due to atmospheric oxidation.

Earlier work

The basis for this maturity indicator was presented in a paper by Chung and Sackett (5). Their data for 19 different coal samples and 12 different shale samples are reproduced in (Figure 2). For this diverse suite of samples a systematic trend is seen from immature shales in the upper right through lignites in the center to mature shales and anthracites at the lower left.

The exhaustive pyrolysis procedure used by them was to heat samples at 500°C for up to eleven days so that comparisons could be made with earlier work by Sackett and co-workers on model compounds. The most extensive application of the PCI method to date was on a suite of samples of the Bakken shale from the Williston Basin in North Dakota by Conkright et al (6). Their procedure was to pyrolyze for five days at 600°C. They showed a good correlation between CH_4/C and $\Delta^{13}C$ and other maturity parameters.

Problems and questions

Pyrolysis Duration. As mentioned above, Chung and Sackett (5) used eleven days at 500°C and Conkright et al (6) used five days at 600°C These lengths of time are unacceptable for most users who are interested in a quick and practical maturity indicator. As there is no a priori reason not to go to higher temperatures and shorter

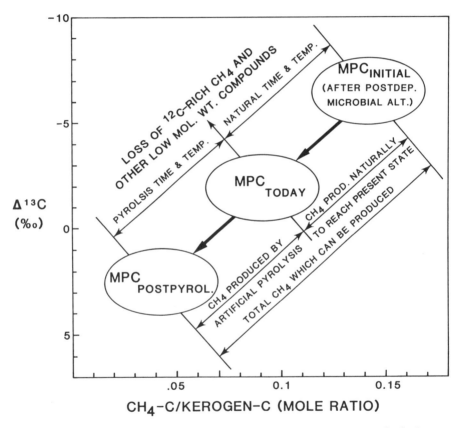

Figure 1.   Model for the Pyrolysis-Carbon Isotope Method for
determining kerogen maturity. MPC is methane precursor carbon.

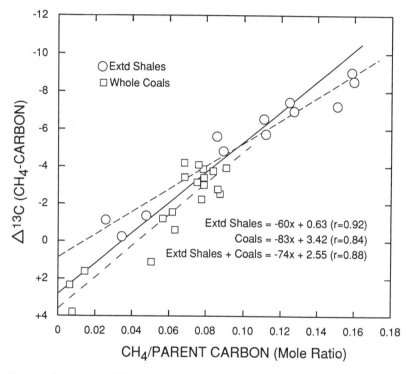

Figure 2. The difference between carbon isotope compositions of pyrolyzed-derived methane and kerogen carbon versus the mole ratio of methane to kerogen carbon for extracted shales and whole coal sample for exhaustive pyrolyses at 500°C.

times our most recent work uses 700°C for one hour. (Table I) presents a time series for pyrolysis of a Bakken shale sample at 700°C. The data show that the amount of methane peaks before one hour, does not change significantly for about 40 hours and then may decrease somewhat for longer times. As will be discussed later this decrease is apparently due to the breakdown of methane. It appears that one hour pyrolysis at 700°C is adequate for exhaustive pyrolysis without fear of methane decomposition.

TABLE I. Methane produced during the exhaustive pyrolysis at 700°C of a Bakken shale sample as a function of time (adapted from Ref. 1)

Pyrolysis Time (hours)	CH$_4$ (ml/100 mg sample)*
0.25	2.70
	2.60
	2.70
0.50	3.10
1	3.60
3	3.40
17	3.70
20	3.40
43	3.60
91	3.30
138	3.10
265	3.10

* precision is estimated at 0.3 ml/100 mg

Carbon isotope exchange between carbon dioxide and methane. In the study on carbon isotope exchange between methane and amorphous carbon (1), the potential exchange with carbon dioxide was minimized by placing a plug of CaO spatially separated from the sample in the pyrolysis tubes for the purpose of reacting with any CO$_2$ to produce inert CaCO$_3$. As the vapor pressure of CO$_2$ over CaCO$_3$ at 700°C is about 2 cm (according to the CRC Handbook of Chemistry and Physics) some CO$_2$ may be present in our pyrolysis tubes.

In order to confirm that isotope exchange with carbon dioxide is not a problem, a series of experiments was run with a 50-50 mixture of methane and carbon dioxide (δ^{13}C-CO$_2$ = -3.7 and δ^{13}C-CH$_4$ = -43.7 ‰ vs PDB). The gas mixture was sealed in glass and heated for various times and temperatures. If isotopic exchange takes place, one would expect that both compositions would converge with the equilibrium mixture showing that carbon dioxide is somewhat enriched in carbon-13 relative to methane.

According to the theoretical work of Bottinga (7) for equilibrium at 700°C, the carbon in carbon dioxide would be 10 o/oo heavier than the carbon in methane. The definitive results shown in (Table II) indicates that at 600°C there is little or no change over 22 days whereas at 800°C there is a significant change in methane but only a slight change in carbon dioxide over a two day period. This work is continuing but strongly suggests minimal carbon isotope exchange over one hour at 700°C and that methane is breaking down over longer time periods with a large fractionation between it and its product(s).

TABLE II. Data for carbon isotope exchange between methane and carbon dioxide at 600 and 800°C

Experiment	Chemical Species	Carbon Isotope Exchange at 600°C Time (hours)	$\delta^{13}C*$
1.	CO_2	0	-4.0
	CH_4		-44.1
2.	CO_2	0	-3.6
	CH_4		-43.5
3.	CO_2	120	-4.4
	CH_4		-45.4
4.	CO_2	122	-3.8
	CH_4		-44.8
5.	CO_2	524	-4.2
	CH_4		-45.2
6.	CO_2		-3.9
	CH_4		-45.2
7.	CO_2		-3.8
	CH_4		-44.6
Carbon isotope Exchange at 800°C			
1.	CO_2	0	-3.7
	CH_4		-43.8
2.	CO_2	18.5	-2.6
	CH_4		-28.4
3.	CO_2	43	-4.8
	CH_4		-23.2
4.	CO_2	44.5	-4.7
	CH_4		-26.6

* in ‰ versus PDB.

Methane stability at high temperatures. Several of the experiments discussed above suggest that methane may decompose at temperatures above 600°C. To determine the threshold temperature for this to happen, pure methane was sealed in Vycor tubes and placed in an oven at 700, 800, 900 and 1000°C. Data are given in (Figure 3). They suggest that at temperatures of about 700°C methane begins to show some decomposition over a period of several days. At higher temperatures decomposition precedes more rapidly. For our standard time and temperature of one hour at 700°C, little methane decomposition should occur.

Recent Application

The Miocene Monterey Formation in California is an organic-rich siliceous shale that is the source and reservoir rock of major

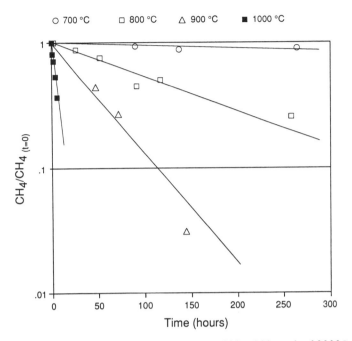

Figure 3. Methane stability at 700, 800, 900 and, 1000°C.

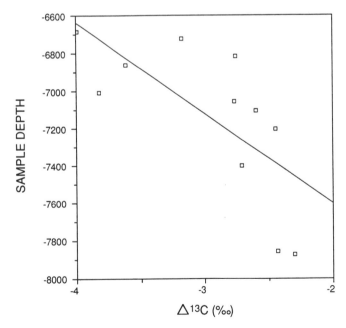

Figure 4. $\Delta^{13}C$ versus core depth (feet) for Monterey rock samples.

hydrocarbon accumulations. The organic matter in the rocks is generally considered to be Type II marine algal material, having good oil-bearing potential. Hydrocarbons from the Monterey Formation are generally heavy with an API gravity less than 20° and are considered unusual because they appear to have been generated from kerogens of low maturity as indicated by kerogen vitrinite reflectance (Ro) values of 0.3 to 0.6 % and Rock-Eval Tmax values of 400 to 440°C (8).

The PCI method was applied to a suite of samples from the Chevron B-2 well drilled from the offshore Platform Hermosa located in the southeastern part of the Point Aguello Field approximately 10 km west of Pt. Conception. The core material was made up of fractured chert, porcellanites and dolostones with from 0.3 to 3.4% total organic carbon. The CH_4/C ratio ranged randomly between 0.05 to 0.2. As shown in (Figure 4), however, $\Delta^{13}C$ values show an overall general increase with burial depth as predicted. Similar to Ro and Tmax, as discussed by Walker et al. (9) it appears that the PCI method, on the basis of these initial results, is not a particularly useful maturity indicator for Monterey rocks.

Conclusions

By decreasing pyrolysis time to one hour at 700°C and showing that isotopic exchange between methane and carbon dioxide and the thermal decomposition of methane do not occur for this time and temperature, the pyrolysis-carbon isotope method should become an even more attractive method for the determination of the maturity of kerogen in whole rock samples.

Acknowledgments

Partial support for this study was provided by the following grants: NSF #OCE-9015580, DOE #DE-FG 05-92 ER 1430 and the donors of the Petroleum Research Fund, administered by the ACS.

Literature Cited

1. Sackett, W.M.; Org. Geochem. 1993, 20, 43-45.
2. Sackett, W.M.; Org. Geochem. 1986, 9, 63-68.
3. Sackett, W.M.; Org. Geochem. 1984, 6, 359-363.
4. Sackett, W.M.; Geochim. et Cosmochim. Acta 1978, 42, 571-580.
5. Chung, H.M.; Sackett, W.M.; Geochim. et Cosmochim. Acta 1979, 43, 1979-1988.
6. Conkright, M.E.; Sackett, W.M.; Peters, K.E.; Org. Geochem. 1986, 10, 1113-1117.
7. Bottinga, Y.; Geochim. et Cosmochim. Acta 1969, 33, 49-64.
8. Peterson, N.F.; Hickey, P.J.; In Exploration for Heavy Crude Oil and Natural Bitumen; Studies in Geology 23; American Association of Petroleum Geologists: Tulsa, OK, 1987.
9. Walker, A.L.; McCulloh, T.H.; Petersen, N.F.; Stewart, R.J.; in Petroleum Generation and Occurrence in the Miocene Formation California; (Eds.; C.M. Isaacs and R.E. Garrison), Pacific sec. SEPM: Tulsa, OK, 1983, 185-190.

RECEIVED May 13, 1994

APPLICATION OF VITRINITE REFLECTANCE TO BASIN MODELING

Chapter 14

Calculation of Vitrinite Reflectance from Thermal Histories and Peak Temperatures

A Comparison of Methods

Charles E. Barker[1,2] and Mark J. Pawlewicz[1]

[1]U.S. Geological Survey, Box 25046, MS 971, Denver, CO 80225
[2]University of Adelaide, South Australia 5005, Australia

One purpose of vitrinite reflectance, a thermal maturation parameter, is to characterize the degree of heating in sedimentary rocks. This use of vitrinite reflectance makes its calibration to peak temperature (T_{peak}) important because T_{peak} constitutes an absolute thermal maturation parameter. This paper compares the two major methods available for predicting thermal maturation through vitrinite reflectance evolution: thermal history (kinetic models) and T_{peak} (vitrinite reflectance geothermometers, VRG).

VRG applied to the published cases of burial heating show that the mean random vitrinite reflectance (R_{v-r}), predicted from kinetic models closely agrees with the measured R_{v-r} when a realistic or measured T_{peak} value is used in setting the thermal history. Similarly, a published comparison of these two types of VRG and kinetic models in an active hydrothermal metamorphism (geothermal) system suggests that VRG give predictions that are a better match to present day temperatures.

The purpose of thermal maturation parameters, such as vitrinite reflectance, is to measure the degree of heating. This use of vitrinite reflectance makes its calibration to peak temperature (T_{peak}) important because T_{peak} constitutes an absolute thermal maturation parameter. T_{peak} is also a key inflection point in burial history reconstruction and is a physically significant point with which to compare the organic geochemistry of samples.

Thermal maturation parameters are widely used to correlate with other physical parameters. One of these physical parameters, T_{peak}, is a fundamental physical value rather than a thermal maturation parameter defined by an analytical method, such as mean random vitrinite reflectance (R_{v-r}). Studies using temperature-based comparisons rather than thermal maturation parameters have an obvious advantage in that standards and instrumentation are readily available worldwide and its connection to other physical sciences is quantitative.

T_{peak} is limited in its application to sedimentary basins because direct measurements are possible only in strata that can be demonstrated to be at the highest temperature attained in their geologic history. In rocks that have cooled, T_{peak} can be determined by using fluid inclusions, apatite fission track analysis (AFTA) or measurements on equilibrium mineral assemblages. AFTA studies are limited in that tracks are not generally retained above about $100^{\circ}C$. Studies of equilibrium mineral assemblages are also limited in that during burial heating, equilibration can be difficult to attain or may not be recorded by mineral precipitation. These problems make it advantageous to measure other more commonly available and easily measured thermal maturation parameters. Thus, T_{peak} is presently not considered a replacement, but supplementary, to vitrinite reflectance, AFTA, or other thermal maturation parameters.

The reason that other thermal maturation parameters are needed is because R_{v-r} is subjective in its measurement and its accuracy should be assessed in each case. Subjectivity in R_{v-r} measurements results from a microscope operator selecting the vitrinite particle that has its reflectance measured quantitatively. Thus, the scatter in R_{v-r} data is often attributable to differences in laboratory technique and problems with finding and measuring the same vitrinite group maceral between samples. These operational and methodological factors restrict the precision to one decimal place even though the precision of the physical operation (i.e. microphotometry) is 0.01%. Other physical and chemical changes such as reflectance suppression and pressure can also locally affect vitrinite reflectance evolution. Such problems, although limiting the effectiveness of R_{v-r} in some cases, do not invalidate the general use of such data.

Two major methods are available for predicting thermal maturation through vitrinite reflectance evolution: thermal history (kinetic models) and T_{peak} (vitrinite reflectance geothermometers, VRG). The thermal history method sums, over the entire burial history of a rock package, the extent of R_{v-r} evolution reactions that are modeled using kinetic equations. The T_{peak} method considers that chemical reactions up to peak temperature are dominant in setting the cumulative R_{v-r} evolution. In other words, kinetic models allow time as an unlimited factor and assume that time can be effective in R_{v-r} evolution after marked temperature decreases or after hundreds of millions of years at constant temperature. VRG consider time to be a factor limited in effectiveness to the time as T_{peak} is approached.

We think the most effective way to confirm and compare these methods is by using cases formulated by other scientists. For this reason, this paper focuses on calibration of VRG and their comparison to published kinetic models for determining vitrinite reflectance evolution. The examples used in this comparison are the hypothetical and actual geological case studies of burial heating systems presented by Morrow and Issler *(1)* and Bray et al. *(2)*. For geothermal systems, which are considered as cases of active hydrothermal metamorphism, we use data from Whelan et al. *(3)* for comparison.

Vitrinite Reflectance Geothermometry

Previous Calibrations. Previous calibrations of VRG by us were made using published data, in which T_{peak} was estimated from geologic reconstruction and by grouping burial heating and hydrothermal metamorphism environments together. This grouping of data produced a scattered data set *(4)*   but one that had a surprisingly high coefficient of correlation. The scatter in the calibration of *(4)* obscured any indication that different burial heating and hydrothermal metamorphism environments may follow different thermal maturation paths. However, the point of *(4)* was that there was a significant correlation between T_{peak} and R_{v-r} and that time must have a reduced effect. Much of the scatter in this calibration is now attributed to: 1) differences in R_{v-r} measurements between laboratories *(5)*; 2) grouping equilibrium temperature log data from geothermal systems with uncorrected temperature log data from burial heating cases; 3) the relatively crude estimates of T_{peak} from burial history reconstruction methods; and 4) differences in heating rate and pressure *(6)*. However, this scatter is not due to using random reflectance (R_{v-r}) rather than maximum reflectance at high thermal maturity, because R_{v-r} is strongly correlated with maximum reflectance *(7)* over a wide reflectance range.

Barker and Goldstein *(8)* also grouped burial heating and hydrothermal metamorphism paths together, even though two subtle trends in the data were present, because of the perception that no difference in thermal maturation at a given T_{peak} should be present regardless of the thermal maturation path. The main point of *(8)* was to demonstrate that fluid inclusions can reequilibrate and approach T_{peak} and, as support for this argument, demonstrate a strong relationship with R_{v-r} which is also thought to be controlled by T_{peak}.

The grouping of data from hydrothermal metamorphism and burial heating environments in all of these calibrations produces a curve that predicts temperatures somewhat high for burial heating in the upper temperature ranges. New insights on the effect of heating rate and pressure on vitrinite reflectance evolution suggest it is best to develop regression equations for each environment to produce an optimal calibration for the more accurate prediction (as done in Barker *(9)* and expanded in Aizawa *(10)* and Barker*(11)* rather than force T_{peak} and R_{v-r} data from diverse environments into a single curve as has been done in the past.

The different thermal maturation paths are thought to be due to physical differences in each environment, such as pressure and the ability of the environment to allow escape or removal of reaction products generated during thermal maturation *(6)*. Differences in heating duration and heating rate are seemingly of reduced importance because the systems used in calibrating these paths span a wide range of elapsed time and heating rate. Heating duration is thought to be effective only over a short term reaction period in which the stabilization of organic matter thermal maturation occurs *(11,12)*.

Present Calibration. Calibration of the present VRG is based on a direct estimate of T_{peak}. This estimate can be made by determining the upper mode in a distribution of reequilibrated fluid-inclusion homogenization temperatures (T_h) *(8)*. A minimum estimate of T_{peak} is also possible using the upper limit of the T_h range measured in secondary fluid inclusions. T_{peak} estimated from fluid inclusion data, is more precise because it is a direct temperature measurement made using rocks from the system and is independent of the uncertainty present in burial history reconstruction.

The improved resolution of T_{peak} from fluid inclusions data indicates that vitrinite reflectance evolution during burial heating, hydrothermal metamorphism and contact metamorphism by dikes follow different paths. Hydrothermal metamorphism includes extinct systems, which often formed hydrothermal ore deposits, and active geothermal systems. Contact metamorphism by dikes is discussed by Barker et al. *(13)*, but not further here, as research on this issue is still in progress.

T_{peak} and R_{v-r} data from an updated compilation based mostly on Barker *(9)*, Aizawa *(10)*, and Barker and Goldstein *(8)* and unpublished data are used to calibrate a geothermometer for the burial heating and hydrothermal metamorphism paths. We also group the results based on gross heating rate given by burial heating and hydrothermal metamorphism thermal maturation paths. Reduced major axis regression (Fig. 1; Table I) suggests a calibration curve of :

$$T_{peak} = (\ln(R_{v-r})+1.68)/\, 0.0124 \text{ for burial heating } (r^2 = 0.7; n=51)$$

and,

$$T_{peak} = (\ln(R_{v-r})+1.19)/0.00782 \text{ for hydrothermal metamorphism } (r^2=0.7; n=72).$$

Data scatter about these different thermal maturation paths is attributed to the second order effects of heating rate, pressure and system closure within each environment, measurement error in both variables, and so forth.

Comparison of Kinetic Models and Vitrinite Reflectance Geothermometry

Burial heating. Morrow and Issler *(1)* provided a thorough review of the theoretical basis of kinetic models and such a discussion is not repeated here. Note that in the following discussion the % Ro symbol used in *(1)* for vitrinite reflectance represents the mean random vitrinite reflectance measured in oil (R_{v-r} ,as described above).

Figure 2 shows R_{v-r} calculated by Morrow and Issler *(1, see their figure 2)* for selected kinetic models for a constant temperature persisting over a 200 m.y. reaction duration. Note that VRG predictions are comparable near an elapsed time of about 10 m.y. The VRG predictions are about 10 to 20% lower for an elapsed time of 200 m.y than the predictions of the more conservative kinetic models presented by *(1)*. The low R_{v-r} predictions at very large elapsed times for time-limited models are expected because VRG do not consider increasing R_{v-r} after T_{peak} is reached. This stabilization of vitrinite reflectance evolution is in contrast to kinetic models that theoretically can allow vitrinite reflectance to increase forever.

Figure 3 shows the idealized temperature histories presented by Morrow and Issler *(1; see their figure 5)* along with R_{v-r} values estimated from the burial heating VRG presented in this paper. In both cases, the R_{v-r} estimates based on the VRG show close agreement with those of the kinetic models.

Figure 4 shows the R_{v-r} values computed from an estimate of peak paleotemperature of $117^{\circ}C$ based on apatite fission track analysis (AFTA) on samples from a depth of 805 m in the Esso Bissette well *(1; see their Fig. 6)*. The bar through the R_{v-r} predictions shows the computed range based on the acceptable T_{peak} values quoted by *(1)*. These VRG predictions overlap with the predictions from the kinetic models and the measured R_{v-r}. Note that Morrow and Issler *(1)* conclude that none of the estimates from kinetic models fit the data very well in this reconstruction. Perhaps suppression of vitrinite reflectance is a possible explanation for the poor fit *(19, 20)*.

Table I. Peak Temperature Estimated from Vitrinite Reflectance

Mean Random Reflectance (0.01 % increment)	Burial Heating Tpeak (Celsius)	Hydrothermal Metamorphism Tpeak (Celsius)	Mean Random Reflectance (0.1 % increment)	Burial Heating Tpeak (Celsius)	Hydrothermal Metamorphism Tpeak (Celsius)
0.40	62	35	1.1	144	164
0.41	64	38	1.2	151	175
0.42	66	41	1.3	157	186
0.43	68	44	1.4	163	195
0.44	70	47	1.5	169	204
0.45	71	50	1.6	174	212
0.46	73	53	1.7	179	220
0.47	75	56	1.8	184	227
0.48	77	58	1.9	188	234
0.49	78	61	2	192	241
0.50	80	64	2.1	196	247
0.51	82	66	2.2	200	253
0.52	83	69	2.3	203	259
0.53	85	71	2.4	207	264
0.54	86	73	2.5	210	269
0.55	88	76	2.6	213	274
0.56	89	78	2.7	216	279
0.57	91	80	2.8	219	284
0.58	92	83	2.9	222	288
0.59	93	85	3	225	293
0.60	95	87	3.1	228	297
0.61	96	89	3.2	230	301
0.62	97	91	3.3	233	305
0.63	99	93	3.4	235	309
0.64	100	95	3.5	237	312
0.65	101	97	3.6	240	316
0.66	102	99	3.7	242	319
0.67	104	101	3.8	244	323

0.68	105	103	3.9	246	326
0.69	106	105	4	248	329
0.70	107	107	4.1	250	333
0.71	108	108	4.2	252	336
0.72	109	110	4.3	254	339
0.73	111	112	4.4	256	342
0.74	112	114	4.5	258	345
0.75	113	115	4.6	260	347
0.76	114	117	4.7	261	350
0.77	115	119	4.8	263	353
0.78	116	120	4.9	265	355
0.79	117	122	5	266	358
0.80	118	124	5.1	268	361
0.81	119	125	5.2	270	363
0.82	120	127	5.3	271	365
0.83	121	128	5.4	273	368
0.84	122	130	5.5	274	370
0.85	123	131	5.6	276	372
0.86	124	133	5.7	277	375
0.87	125	134	5.8	278	377
0.88	126	136	5.9	280	379
0.89	127	137	6	281	381
0.90	128	139	6.1	282	383
0.91	128	140	6.2	284	385
0.92	129	142	6.3	285	388
0.93	130	143	6.4	286	390
0.94	131	144	6.5	288	392
0.95	132	146	6.6	289	393
0.96	133	147	6.7	290	395
0.97	134	148	6.8	291	397
0.98	134	150	6.9	292	399
0.99	135	151	7	294	401
1	136	152			

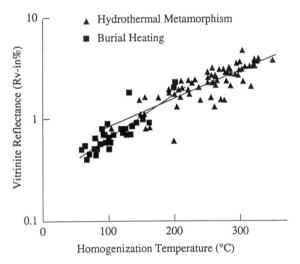

Figure 1. Peak temperature (T_{peak}) from reequilibrated fluid inclusions versus mean random vitrinite reflectance (R_{v-r}) for burial heating and hydrothermal metamorphism. Figure based on Barker & Goldstein (8). Additional data from Aizawa (10) and unpublished data. Hydrothermal metamorphism includes data from hydrothermal ore deposits and active geothermal systems.

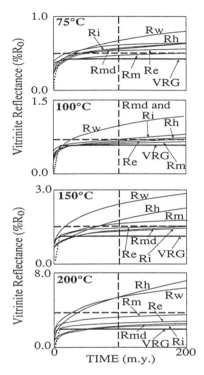

Figure 2. Comparison of predicted R_{v-r} based on Figure 2 of Morrow and Issler *(1)*. Solid lines indicate constant-temperature plots of vitrinite reflectance evolution for the various models presented by Morrow and Issler *(1)* and the VRG for burial heating. R_{v-r} estimates made from Morrow and Issler *(1)* = Ri; Middleton *(14)* = Rmd; Waples *(15)* = Rw; Sweeney and Burnham *(16)* = Re; MATOIL *(17)* = Rm; Horvath et al. *(18)* = Rh; VRG = Burial heating calibration, this paper. Note that some of the more recently published kinetic models predict little significant change in R_{v-r} as time increases after about 20-30 m.y. suggesting stabilization of vitrinite reflectance evolution.

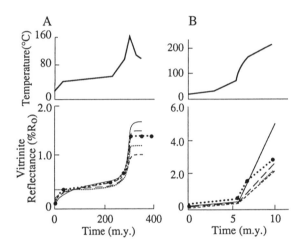

Figure 3. Plots of calculated R_{v-r} in response to idealized temperature histories in tectonic settings with moderate to rapid heating rates. Plot (A) models the temperature history of a foreland basin and plot (B) models temperature history of a wrench fault basin. The lines represent R_{v-r} predictions for models presented by Morrow and Issler *(1, see their Figure 5)* and the VRG for burial heating (this paper). The lines represent (—) Waples *(15)*; (— —) Matoil *(17)*; (····) Easy%R_O *(16)*; (—) Issler *(1)* and Middleton *(14)*; and (•····•) VRG, burial heating, this paper.

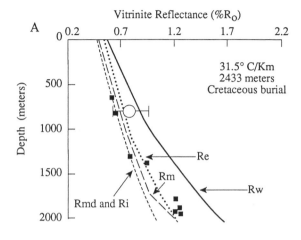

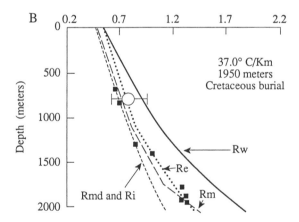

Figure 4. Two burial history scenarios taken from Figure 6 of Morrow and Issler *(1)* for the Esso Bissette well. R_{v-r} data (solid squares) are compared to the calculated R_{v-r} value (circle and range bar) from VRG calibration for burial heating (this paper). The R_{v-r} is calculated using a T_{peak} of 117°C estimated from apatite fission track analysis. The range bar through the R_{v-r} predictions shows the computed R_{v-r} range based on the range of acceptable T_{peak} quoted by Morrow and Issler *(1)*. The lines represent (—) Waples *(15)*; (— —) Matoil *(17)*; (....) Easy%Ro *(16)*; (—) Issler *(1)* and Middleton *(14)*; and (•····•) VRG, burial heating, this paper.

The last comparison in burial heating systems uses Bray et al. *(2)* who present vitrinite reflectance and paleotemperature data from the 47/29a-1 well drilled into the East Midlands Shelf, United Kingdom. The 47/29a-1 well is substituted for the Gilbert F-53 well case study that Morrow and Issler present because T_{peak} in *(1)* is only estimated from geologic models of burial and is not as well known as their other natural examples. Figure 5 shows a comparison of various model predictions to the measured R_{v-r} from the 47/29a-1 well *(2, see their figures 4 and 5)*. VRG predictions of R_{v-r} are in good agreement with the predictions from the Sweeney and Burnham *(16)* kinetic model and paleotemperatures estimated from AFTA.

Hydrothermal Metamorphism in Geothermal Systems. VRG for hydrothermal metamorphism is a well documented and accepted technique *(21, 22, 23)*. A case study by Whelan et al. *(3)* compared the predictions of VRG of *(9)* and kinetic models to oxygen isotope geothermometry and measured present-day temperatures. These measured temperatures are thought to be near T_{peak}. Whelan et al. *(3)* found that predictions from VRG for hydrothermal metamorphism (Fig. 6) are in good agreement with the present temperatures and oxygen isotope geothermometry. They concluded that the predictions from the kinetic model of Sweeney and Burnham *(16)* did not give as good of a fit to present day temperatures as the VRG. Further, the heating rate of $5^{\circ}C/$ m.y. input into the kinetic model of *(16)* gave the best fit to the present day temperatures. This heating rate does not fit well with the known age of these ocean bottom sediments that are estimated at being heated to near $280^{\circ}C$ for most of the past 125,000 years *(3)*.

Discussion

These comparisons show that, based on the hypothetical and natural examples presented by Morrow and Issler *(1)*, as well as those of Bray et al. *(2)* and Whelan et al. *(3)*, VRG predictions agree favorably with predictions from kinetic models where T_{peak} is well known. Thus, Morrow and Issler *(1)* were in error when they arbitrarily excluded VRG from consideration by asserting such models have been contradicted by time-temperature modeling of Burnham and Sweeney *(24)*. In fact, Burnham and Sweeney *(25)* and Sweeney and Burnham *(16, 25)* found that predictions from VRG and kinetic models compare favorably. Similar results have been found by *(26, 27)* and Laughland, this volume, and many others who have direct measurements of T_{peak} to make the VRG and kinetic model predictions. The kinetic model of *(16)* also suggests negligible changes in R_{v-r} of less than 0.1% after some 20 to 30 m.y. at a constant (peak) temperature (Fig. 2). This is a negligible change in R_{v-r} because microscopist and laboratory bias and other possible errors in reflectance measurements lead to the general observation that reflectance values are limited in precision to one decimal place *(28)*. Further, Barker and Crysdale *(29)* found that R_{v-r} value predicted from the Sweeney and Burnham *(16)* kinetic model after adjusting the thermal history to fit T_{peak} closely agrees with the measured R_{v-r}. Thus, the kinetic model of *(16)* supports the concept of stabilization after about 20 to 30 m.y. This stabilization of vitrinite reflectance evolution is the theoretical basis of VRG. We also point out that Morrow and Issler *(1)* misstate the theoretical basis of VRG by mingling the observations of Price *(30)* on the importance of higher order reactions with the observations of Barker and Pawlewicz *(4)* that R_{v-r} is strongly correlated with T_{peak}. These two observations are separate and are not linked by us.

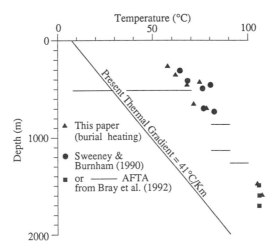

Figure 5. A comparison of measured geothermal gradient and T_{peak} estimates in the 47/29a-1 well, East Midlands Shelf, United Kingdom. Modified from Bray et al. *(2, see their figures 4 and 5)*. The measured geothermal gradient suggests the well has cooled from T_{peak}. Note VRG predictions (solid triangles) are in good agreement with the paleotemperature predictions from the Sweeney and Burnham *(16)* kinetic model (dots) and AFTA (solid squares or a horizontal line which indicates the estimated paleotemperature range).

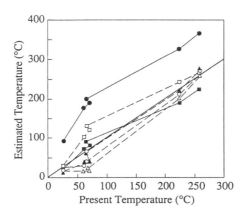

Figure 6. Whelan et al. *(3)* comparison of various geothermometry and predictions from the Sweeney and Burnham *(16)* kinetic model for various heating rates. The solid (y=x) line represents perfect agreement between temperature estimate and measurement. All computations, present temperature measurements and paleotemperature estimates from Whelan et al. *(3)*, except for the geothermometry for hydrothermal metamorphism (from this paper). For the purposes of plotting, the data values of *(3)* marked less than or greater than were rounded down or up by $5°C$, respectively. The data symbols: (o) O-isotopes geothermometry *(3)*; (▲) Hydrothermal metamorphism this paper; (Δ) Geothermal *(9)*; (□) 1 °C/1,000 yr *(16)*; (●) 10 °C/yr *(16)*; and (■) 5 °C/my *(16)*.

Barker and Pawlewicz *(4)* empirical observation infers that reaction time does not have a continuing effect on R_{v-r} and thus, vitrinite reflectance evolution must stabilize at T_{peak} after some limited reaction time period. Barker *(11)* suggested that stabilization of vitrinite reflectance evolution in burial heating systems requires up to 10 m.y. — this is not a geologically insignificant period as Morrow and Issler *(1)* state. The R_{v-r} predicted by the more conservative of the kinetic models that Morrow and Issler present (Fig. 2) show that, in practical terms, there is little discernible change in vitrinite reflectance past 20-30 m.y. This is a small difference in the estimated stabilization time given the simplifications inherent in both methods.

Concluding Remarks

VRG have proven accurate or the only applicable method in: 1) accretionary or orogenic tectonic regimes where the burial history or heating is complex or reconstructible only with considerable uncertainty *(27, 31, among others)*; 2) in cases where direct measures of T_{peak} are available *(26, 32, 33)*; and 3), in geothermal systems *(21, 22, 23)*.

We acknowledge that in geological systems where the burial history is well known, kinetic methods are accurate especially if heating rate can be measured and applied to the proper kinetic model to refine the T_{peak} prediction. We recommend that the method used should reflect the quality of the data that is available in a geologic system. If the thermal history and(or) burial history is poorly known, the use of VRG seems appropriate because it needs only a R_{v-r} measurement to make an estimate of T_{peak}. We reiterate, however, that in our studies where the burial history is well known *(as in 30)*, we have found that if the burial history is modified to consider a T_{peak} measured in that system, comparable vitrinite reflectance predictions can be made with both VRG and kinetic methods.

Acknowledgments

This paper was much improved and focused by the reviewers. We thank B. J. Cardott of the Oklahoma Geological Survey, and J. L. Clayton, M.A. Henry, D.K. Higley, J. D. King and M.D. Lewan of the U.S. Geological Survey.

Literature Cited

1. Morrow, D.W.; Issler, D.R., *Am. Assoc. Pet. Geol.* **1993**, *77*, 610-624.
2. Bray, R.J., Green, P.F., and Duddy, I.R., In *Exploration Britain: Geological Insights for the Next Decade*; Hardman, R.F., Ed., Special Publication no. 67; Geological Society: London, **1992**, 3-25.
3. Whelan, J.K., Seewald, J., Eglinton, L. and Miknis, F. *Initial Reports,* Ocean Drilling Program. in press.
4. Barker, C.E., and Pawlewicz, M.J. In *Paleogeothermics;* Buntebarth, G., and Stegena, L., Ed.; Springer-Verlag New York, 1986, pp. 79-93.
5. Dembicki, H., Jr. *Geochim. Cosmochim.,* **1984**, *48*, 2641-2649.
6. Teichmüller, M., In *Coal and Coal-bearing Strata: Recent Advances;* Scott, A.C., Ed., Special Publication 32; Geological Society: London, 1987, 127-169.
7. Kilby, W.E. *Int. J. Coal Geol.,* **1991**, *19*, 201-218.
8. Barker, C.E., and Goldstein, R.H. *Geology,* **1990**, *18*, 1003-1006.

9. Barker, C.E. *Geology*, **1983**, *11*, 384-388.
10. Aizawa, J. In *Proceedings;* International Conference on Coal Science, **1989**, *1*, 93-96.
11. Barker, C.E. In *Thermal History of Sedimentary Basins*; Naeser, N.D., and McCulloh, T.H., Eds.: Springer-Verlag, New York, 1989; pp. 73-98.
12. Barker, C.E. *Am. Assoc. Pet. Geol. Bull.*, **1991**, *75*, 1852-1863.
13. Barker, C.E., Bone, Y., Duddy, I.R., Marshallsea, S.J., and Green, P.F. *Abstracts Soc. Org. Petrol.*, **1992**, *9*, 102-104.
14. Middleton, M.F. *Geophys. J. Royal Astro. Soc.* **1982**, *68*, 121-132.
15. Waples, D. *Am. Assoc. Pet. Geol. Bull.*, **1980**, *64*, 916-926.
16. Sweeney, J.J., and Burnham, A.K. *Am. Assoc. Pet. Geol. Bull.*, **1990**, *74*, 1559-1570.
17. BEICIP, *MATOIL, release no. 2* : Bureau d'Etudes industrielles et de Cooperation de l'Institut Francais du Petrole, Paris, 1990.
18. Horváth, F., Dövényi, P., Szalay, Á., and Royden, L.H. In *The Pannonian Basin—a Study in Basin Evolution*; Royden, L.H., and Horváth, F., Eds., Memoir 45, Am Assoc Pet. Geol., **1988**, 355-372.
19. Wilkins, R.W.T., Wilmshurst, J.R., Russell, N.J., Hladky, G., Ellacott, M.V., and Buckingham, C. *Org. Geochem.*, **1992**, *18*, 629-640.
20. Lewan, M.D. *Abstracts Soc. Org. Petrol.*, **1993**, *10*, 1-3.
21. Taguchi, K., Mitani, M., Inaba, T. and Shimada, I. In *Proceedings,* Fourth International Symposium on Water-Rock interaction: 1983, 455-458.
22. Gonsalez, R.C. *Vitrinite Reflectance of cores and cuttings from the Ngatamariki well NM-2*; Report 85.10; New Zealand Geothermal Institute: Auckland, 1985, 60p.
23. Struckmeyer, H., and Browne, P.R.L., In *Proceedings;* Tenth New Zealand Geothermal Workshop. **1988**, 251-255.
24. Burnham, A.K., and Sweeney, J.J. *Geochim. Cosmochim*, **1989** *53*, 2649-2657.
25. Sweeney, J.J., and Burnham, A.K. *Am. Assoc. Pet. Geol. Bull.*, **1993**, 77, p. 665-667.
26. Tilley, B.J., Nesbitt, B.E., and Longstaffe, F.J. *Am. Assoc. Pet. Geol. Bull.*, **1989**,*73*, 1206-1222.
27. Laughland, M.M. *Organic metamorphism and thermal history of selected portions of the Franciscan accretionary complex of Coastal California*; unpublished Ph.D. thesis, University of Missouri: Columbia, MO, 1991, 318 p.
28. Robert, P. *Organic Metamorphism and Geothermal History*; D. Reidel Publishing Co.: Boston. MA, 1988, 311p.
29. Barker, C.E., and Crysdale, B.L. In *50th field conference guidebook;* Stroock , B. and Andrew, S., Eds.; Wyoming Geological Association: Casper, WY, 1993; 235-258.
30. Price, L.C. *J. Pet Geol.*, **1983**, *6*, 5-38.
31. Yang, C., and Hesse, R. *Org. Geochem.*, **1993**, *20*, 381-403.
32. Bone, Y., and Russell, N.J. *Aust. J. Earth Sci.*, **1988**, *35*, 567-570.
33. Diessel, C.F.K., *Coal-Bearing Depositional Systems*; Springer-Verlag: New York, NY, 1992; 661p.

RECEIVED April 15, 1994

Chapter 15

Measured Versus Predicted Vitrinite Reflectance from Scotian Basin Wells

Implications for Predicting Hydrocarbon Generation–Migration

P. K. Mukhopadhyay[1], J. A. Wade[2], and M. A. Williamson[2]

[1]Global Geoenergy Research Limited, 14 Crescent Plateau, Halifax, Nova Scotia B3M 2V6, Canada
[2]Geological Survey of Canada, Atlantic Geoscience Centre, P.O. Box 1006, Dartmouth, Nova Scotia B2Y 4A2, Canada

Model predicted vitrinite reflectance data are calibrated with measured reflectance profiles for wells in the Scotian Basin, Eastern Canada. The predicted data were derived using a basin modeling package (BasinMod) which assumed a rifting heat flow at 200 Ma, a variable stretching factor (β), and present day heat flow of 42 mW/m^2. Using combined measured and predicted maturity parameters, measured kinetics on selected source rock kerogens (Kerogen Type IIA-IIB and IIB), and a kinetic model, the timing of hydrocarbon generation and migration from various source rocks (especially Type IIA-IIB from the Lower Cretaceous in N. Triumph G-43 and Type IIB from the Upper Jurassic in Venture B-52) are estimated. Variations in maturation profiles are correlated to abnormal heat transfer due to changes in thermal conductivity or the presence of older rocks (closer to rifting). Anomalies in maturation boundaries are attributed to high sedimentation rates and differences in thermal conductivities due to variations in lithology. Lower Cretaceous (Naskapi Member) source rocks generated and migrated oils during the last 25 Ma; Upper Jurassic-Kimmeridgian Type IIB source rocks migrated mainly condensates and gases between 85-50 Ma.

Present day geochemical basin modeling can illustrate the thermal history and hydrocarbon generation of a source rock through time using a chemical kinetic expression (1-6). Generally, it implies various first order simultaneous reactions with variable activation energies and reaction rates instead of a single reaction as envisaged by empirical basin modeling (7-9). A basin's thermal history will control the generation of hydrocarbons from suitable source rocks. The thermal history is assumed to be derived through convection and conduction (possibly by stretching and rifting of crust) from the upper mantle/lower crust to upper crust (10, 11).

In the Scotian Basin of Eastern both empirical and kinetic basin modeling

0097–6156/94/0570–0230$08.00/0
Published 1994 American Chemical Society

concepts were used by earlier workers to define the level of organic maturation of various sequences through time (9, 12 and 13) and to define hydrocarbon generation rates. Those concepts also define the relationship between overpressuring and gas generation (14 and 15). Since 1989, extensive research on various source rocks, crude oils and condensates taken from several Scotian Basin formations have defined the specific kerogen types and maturation; their distribution in the basin; oil-source rock correlation; and possible hydrocarbon migration (16, 17, and 18). In this study, a commercial basin modeling package (BasinMod; 19) was used to illustrate the following: (a) the influence of various geological parameters affecting thermal modeling or maturation history through time on individual source rocks; (b) the causes of variable maturity in the basin; (c) the relationship between measured kinetics of individual source rocks and the boundary conditions of oil and gas generation; and (d) the possible timing of hydrocarbon migration in relation to their thermal maturity.

Boreholes, Data Input and Limitations of the Data

Eight boreholes (Alma F-67, Cohasset D-42, Evangeline H-98, N. Triumph G-43, S. Desbarres O-76, S. W. Banquereau F-34, Thebaud C-74, and Venture B-52) were selected from a variety of depositional settings for one-dimensional basin modeling (Figure 1). All wells are located on the Scotian Shelf near Sable Island. Figure 2 illustrates the approximate position of the 8 wells within the stratigraphic column of the central Scotian Basin. This present paper focusses on three of the wells, Cohasset D-42, N. Triumph G-43 and Venture B-52, which encountered key source rocks for light oils or condensates. Cohasset D-42 penetrated the Abenaki carbonate platform; Venture B-52 encountered the Missisauga and Mic Mac formations in shallow marine to non-marine deltaic facies; and N. Triumph G-43 sampled Lower Cretaceous and Upper Jurassic distal shelf to basinal facies.

   A series of input data, such as geological parameters, kerogen type, total organic carbon content, compaction, etc., are required to perform the modeling analysis. The geological parameters used follow Wade and MacLean (20). The absolute age ascribed to each formation is referenced to Palmer (21) and Wade and MacLean (20). Lithology, top depth and thickness of each formation were based on CanStrat log analysis and seismic stratigraphy. The Breakup Unconformity or the time of rifting was placed at 200 Ma (22). Dehler and Keen (22) also indicated three phases of rifting: one at 220 Ma, another at 200 Ma and the third at 180 Ma. The data below total depth in the modeled wells were extrapolated from regional geological and geophysical studies. The model is calibrated to measured vitrinite reflectance data for each well. Various default parameters such as decompaction factor, expulsion efficiency, β, heat flow, etc., which were used for the calculation of maturity and hydrocarbon generation, are described in Appendix 1. Table 1 illustrates model calculated conductivity and heat capacity data for standard lithologies such as sandstone, shale etc., and for mixed lithologies.

   The kerogen type and TOC (wt %) values were derived partly from the measured analytical data and partly from an approximation based on lithologic variation, depositional environment and knowledge of Scotian Basin source rocks. Kinetics data for Kerogen Type I, II (considered as Kerogen Type IIA) and III are taken from the model default values. The model allows for mixed kerogen types and,

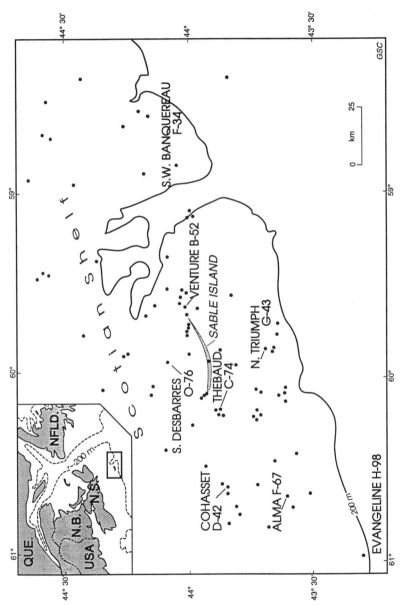

Figure 1. Index map of wells in the Sable Island area of the Scotian Shelf, Canada. Black dots show locations of exploratory and delineation wells. Named wells were modeled for this study.

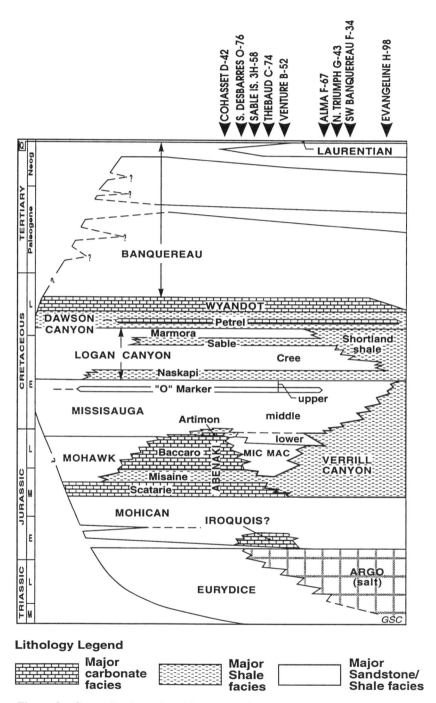

Figure 2.· Generalized stratigraphic column for the central Scotian Shelf. The approximate basin position of the modeled wells is shown at the top.

Table 1

Some examples of matrix conductivity and heat capacity of major sedimentary rocks and mixture of sedimentary rocks.

Lithology Name	Matrix Conductivity (W/m·deg C)	Heat Capacity (KJ/m^deg C)
Pure Rock		
Sandstone	4.40	2800
Siltstone	2.00	2650
Shale	1.50	2100
Limestone	2.90	2600
Dolomite	4.80	2600
Evaporite	5.40	1750
Coal	0.30	950
Igneous Rock	2.90	2500
Mixed Rock Sst/Shale/Lst (*)		
90/10/0	4.11	2730
60/20/20	3.50	2620
50/50/0	2.95	2450
40/50/10	2.800	2430
10/80/10	1.930	2220
0/20/80	2.620	2500
10/30/60	2.600	2470
Mixed Rock Sst/Shale/Salt (*)		
10/10/80	4.910	1890
0/5/95	5.205	1767

(*) sst = sandstone and lst = limestone

therefore, measured activation energies for primary and secondary cracking and reaction constants of Kerogen Type IIA-IIB and IIB were introduced from three laboratory pyrolysis experiments (26). In Venture B-52, the kinetics data of a matured Kerogen Type IIB were introduced. Also, the default density of oil was changed to 0.75 for Kerogen Type IIA-IIB, IIB and III and gas to 0.45 gm/cm³ for Kerogen Types I, II, IIA-IIB, IIB, and III (Hunt, personal communication, 1993). A density of 0.45 gm/cm³ is considered to be the density of gas at the present geological situation and not at the surface. The assumed gas density considers the presence of overpressuring and mixing of condensate with gas at depth. The values used for primary oil and gas potentials are 650 and 70 mg HC/g TOC for Kerogen Type I, 450 and 50 mg HC/g TOC for Type II, 350 and 50 mg HC/g TOC for Type IIA-IIB, 200 and 50 mg HC/g TOC for Type IIB, and 50 and 50 mg HC/g TOC for Type III, respectively (Hunt, personal communication, 1993).

Geology, Organic Facies, Source Rock Types and Their Possible Relationship to Light Oil and Condensate Occurrences

The Scotian Basin developed as a result of the breakup of Pangaea and the formation of the central north Atlantic Ocean during the early Mesozoic. The Triassic-Lower Jurassic synrift sediments are mainly fluvial, lacustrine and aeolian clastic facies of the Eurydice Formation which are gradationally overlain by salt dominated facies of the Argo Formation. The initial post-rift rocks, overlying the Breakup Unconformity at 200 Ma, include clastic and dolostone facies of the Mohican and Iroquois formations respectively. These are, in turn, overlain by a variety of fluvio-deltaic to marine facies of Middle to Late Jurassic age such as the sandstone, shale and minor limestone encountered in the Mic Mac Formation in Venture B-52; the carbonate bank of the Abenaki Formation found in Cohasset D-42; and basinal shale facies of the Verrill Canyon Formation in Alma F-67 and S.W. Banquereau F-34 (Figure 2). The Early Cretaceous was dominated by two thick clastic wedges: the regressive deltaic Missisauga and the transgressive/regressive Logan Canyon formations. The Naskapi Member is a transgressive shale facies of the Logan Canyon Formation. Both the Missisauga and Logan Canyon formations have basinal shale equivalents; the Verrill Canyon Formation and the Shortland Shale (Figure 2). The Logan Canyon Formation is overlain by Upper Cretaceous deeper water marine shale, the Dawson Canyon Formation and the Wyandot Formation chalk. Sandstones and shales of the Tertiary Banquereau Formation complete the stratigraphic succession in the Scotian Basin.

Of specific interest in this study are the light oils, which are reservoired at the top of the Missisauga Formation and in the Naskapi and Cree Members of the Logan Canyon Formation in the Cohasset D-42 well, and the gas/condensates from the top of the Mic Mac and lower member of the Missisauga formations in the Venture field. Sedimentological and source rock analyses indicate that there were only occasional periods of restricted circulation in the basin, from Middle Jurassic through Early Cretaceous, which resulted in limited amounts of Type IIA (classical Type II) (23) oil-prone source rock (SR). Detailed analysis of organic facies suggests that interlaminations of Type IIA-IIB (oil and condensate), Type IIB (condensate and gas), and Type III (gas) source rocks are common (16-18). Using aromatic biomarkers and pyrolysis of asphaltenes, oil-source rock correlation indicates that

Venture condensate correlates with the Type IIB SR in the Mic Mac Formation in the Venture wells (17, 18). The source of the Cohasset oil remains elusive but a source rock similar to Type IIA-IIB algal SR, from the Naskapi Member in the N. Triumph G-43 well, may have made a contribution. Selected samples from the Naskapi source rock contain more than 2.5% TOC content (HI = 270 mg HC/g TOC) and the source rock samples from the upper part of the Mic Mac Formation contain more than 2.0% TOC (HI = 69 mg HC/g TOC). The modeling of these source rocks were used to quantify the amount and timing of hydrocarbon generation in this part of the Scotian Basin.

Results and Discussion

Thermal and maturation models. In order to predict levels of organic maturation versus depth for the wells under study, we constructed a simple, steady-state thermal model which, given an assumed heat flow-time curve and thermal conductivity profile, allows the computation of the location's thermal gradient through time. This forms the basis for estimates of maturity in terms of vitrinite reflectance (see later and appendix). Our preferred heat flow-time assumption is based on a calculated present day heat flow value of 42 mW/m^2 (ie. the product of the actual, measured present day thermal gradient and the thermal conductivity profile) and application of a beta factor of 3.0. A beta factor of 3.0 is, according to Beaumont and Keen (24), an appropriate value to use given the position of the wells on the margin. The maximum rifting induced heat flow was assumed to be at 200 Ma. This declined exponentially to present day values as did its influence on the maturity of sediments deposited since the Early Jurassic.

Our thermal model was translated into estimates of vitrinite reflectance using the Easy %Ro method (4). Figure 3 shows heat flow versus time curves constructed using present day heat flow assumptions of 35, 42 and 48 mW/m2 and a beta factor of 3.0. The influence of these heat flow-time curves on the quality of fit between predicted and measured values of %Ro is shown on Figure 4. For our preferred thermal model, which assumes a present day heat flow of 42 mW/m2, changing the beta factor to 2.5 or 3.5 has no significant impact on the quality of fit (Figure 5) which implies that the effect of the rift induced thermal anomaly dissipated rapidly and has not influenced the maturity evolution of sediments deposited later as seen in Figure 6.

Figures 7 and 8, show the burial histories of the Venture and Cohasset locations together with their computed organic maturity zones. As expected, the higher maturity zones are structurally shallower immediately following the rifting event at 200 Ma. The zones become structurally deeper through time in response to the dissipation of the rifting induced heat flows. A rapid deepening of maturity zones at the Venture B-52 location can be seen at 100 Ma. This is an artifact of the steady state thermal model, as opposed to a transient model, which would smooth out a basin's thermal response and more accurately account for inertia in the subsiding system.

Comparing the burial histories and both the measured and calculated maturity trends in the Cohasset D-42 and Venture B-52 wells, it is observed that Cohasset D-42 shows higher maturity at any given depth (Figures 7, 8 and 9, 10, and 11). The modeled present heat flow for Cohasset D-42 is 42 compared to 40 mW/m^2 in the

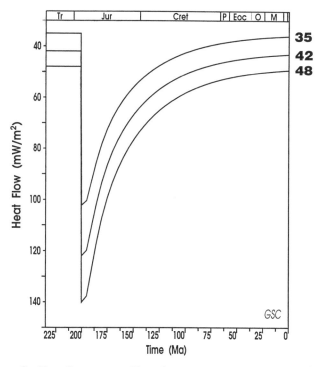

Figure 3.   Heat flow versus Time for Venture B-52, assuming β=3.0.

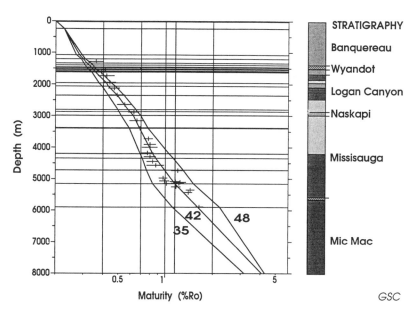

Figure 4.   Measured Ro (+) versus calculated maturity (solid lines) for Venture B-52 using heat flows of 35, 42 and 48 mW/m². For lithology legend, see Figure 11.

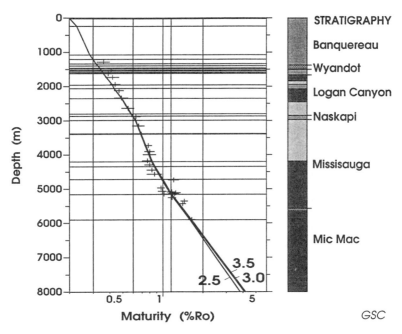

Figure 5. Measured Ro (+) versus calculated maturity (solid lines) for Venture B-52 using β factors of 2.5, 3.0 and 3.5. For lithology legend, see Figure 11.

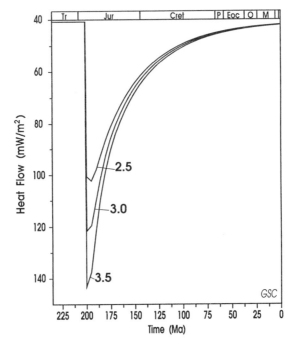

Figure 6. Heat flow versus Time, for β factors of 2.5, 3.0 and 3.5, in Venture B-52 assuming heat flow = 42 mW/m².

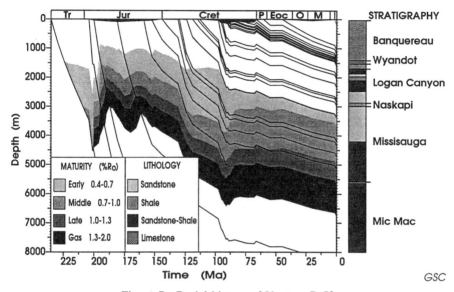

Figure 7.   Burial history of Venture B-52.

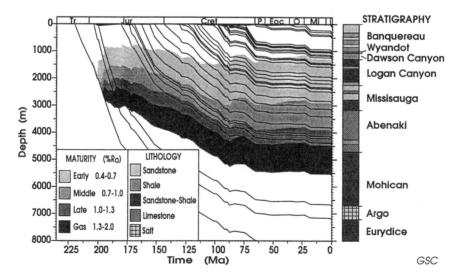

Figure 8.   Burial history of Cohasset D-42.

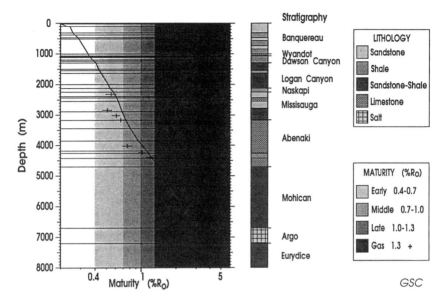

Figure 9. Plot of Maturity versus Depth for Cohasset D-42.

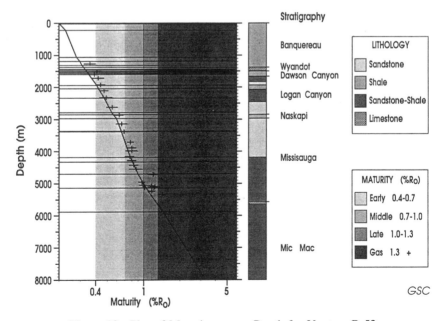

Figure 10. Plot of Maturity versus Depth for Venture B-52.

Venture B-52 to match the measured maturity. Figure 11 shows Cohasset D-42 has higher maturity, especially at the lower part of the well, than Venture B-52 (R_o = 1.2% at 4386 m compared to R_o = 1.22% at 5122 m). This is due to: (a) major differences in lithology and hence differences in thermal conductivity. Cohasset D-42 has thick Jurassic carbonate facies compared with delta front and prodelta (sandstone and shale) sequences in the Venture B-52 well and/or (b) the presence of older rocks at shallower depth (173.6 Ma at 4258 m) in Cohasset D-42 compared to Venture B-52 (144 Ma at 4325 m); the older rocks have experienced higher heat flow due to rifting at 200 Ma.

Measured Kinetics of Source Rocks. Figure 12 shows the distribution of measured activation energies related to the percentages of hydrocarbon generated and their corresponding Arrhenius Constant from three microscopically pre-selected source rock samples. The Kerogen Type IIA-IIB sample, which contains mainly biodegraded algae or amorphous liptinite and has a hydrogen index of 322 mg HC/g TOC, has more than 52% of its hydrocarbons converted below an activation energy of 43 KCal/Mole. All three source rock samples have different activation energies and Arrhenius Constants. The activation energies and Arrhenius Constant values of these samples are comparable to kinetics data of Monterey (Miocene, California, USA) Phosphoria (Permian, Montana, USA), Alum (Cambrian, Sweden) and Woodford (Devonian-Mississippian, Oklahoma, USA) shales determined by Hunt et al. (25) using hydrous pyrolysis. The reaction rates for hydrocarbon generation from Type IIA-IIB kerogen (N. Triumph G-43) are considered as medium-fast similar to those for the Phosphoria Shale. These values are different from the default values of Type I, II and III kerogens as used in the BasinMod program.

Hydrocarbon Generation and Migration.   The timing of crude oil and gas generation and expulsion was quantified by comparing the history of maturation, and the temperature and transformation ratio of individual source rock units in various boreholes. In this study, source rock units from the early Aptian Naskapi Member and Kimmeridgian upper Mic Mac Formation were addressed (Figures 13, 14, 15 and 16). As discussed earlier, the Naskapi Member in the N. Triumph G-43 well has a Kerogen Type IIA-IIB source rock which shows some characteristics similar to the light oils of the Cohasset D-42 discovery. However, it is not the actual source rock which is correlated with the Cohasset D-42 light oils during oil-source rock correlation by aromatic biomarkers. The Kimmeridgian source rock of Venture B-52, however, shows close similarity with the Venture condensate reservoired in the overlying lower member of the Missisauga Formation.

In N. Triumph G-43, the Naskapi source rock had a maturation level of slightly higher than 0.7% R_o. The burial history data show that the Naskapi Member is below 3600 m depth, has been subjected to around 100°C (heat flow around 75 mW/m^2) for the last 35 Ma (Figures 17 and 18) and has a transformation ratio more than 0.8 (26). Figures 15 and 18 show the maturation and temperature history of Naskapi source unit through time. Figure 15 also shows the cumulative generation and migration of liquid and gaseous hydrocarbons from this source rock. Accordingly, more than 200 barrels/acre ft of crude oil have been generated during the last 100 Ma and about 25-30 barrels/acre ft have been migrated during the last 25 Ma. This is due to low activation energies of Kerogen Type IIA-IIB.

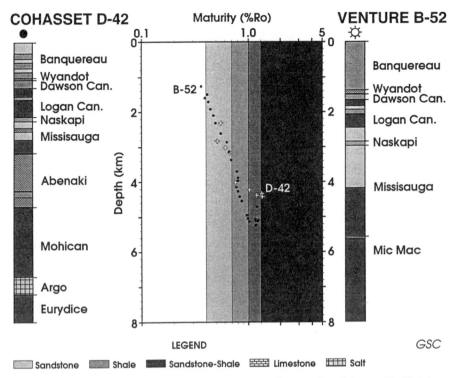

Figure 11. Comparison of Maturity versus Depth plots for Cohasset D-42 (+) and Venture B-52 (•).

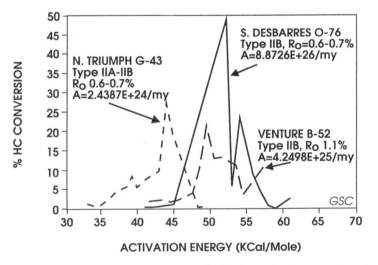

Figure 12. Kinetics data for three source rocks having activation energies related to % hydrocarbon conversion.

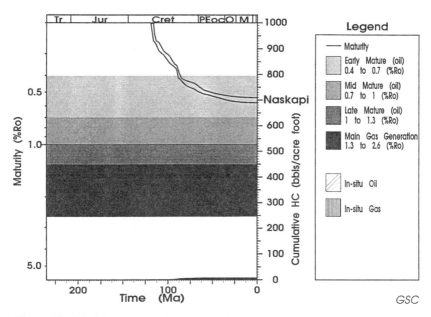

Figure 13. Modeled hydrocarbon generation and maturation of Type IIB source rock from the Naskapi Member of the Logan Canyon Formation in Cohasset D-42.

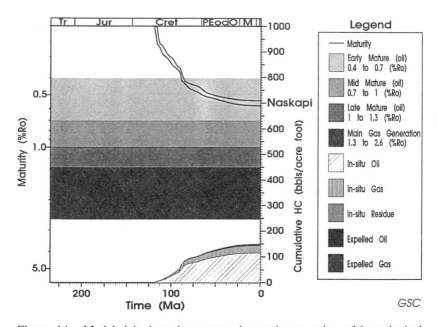

Figure 14. Modeled hydrocarbon generation and maturation of hypothetical Type IIA-IIB source rock from the Naskapi Member of the Logan Canyon Formation in Cohasset D-42.

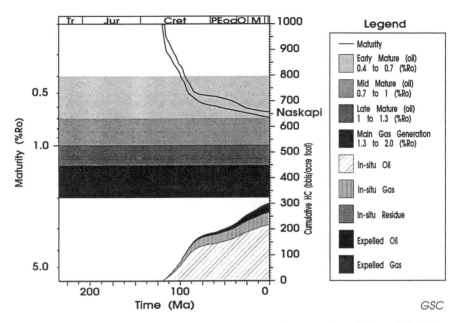

Figure 15. Modeled hydrocarbon generation and maturation of Type IIA-IIB source rock from the Naskapi Member of the Logan Canyon Formation in North Triumph G-43.

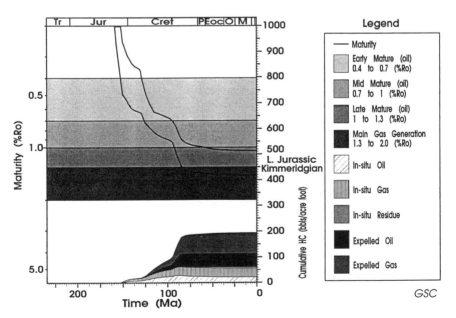

Figure 16. Modeled hydrocarbon generation and maturation of Type IIB source rock from the top of the Mic Mac Formation in Venture B-52.

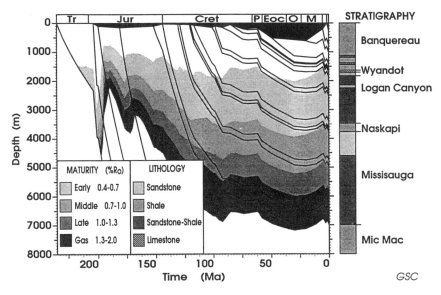

Figure 17. Burial history of North Triumph G-43.

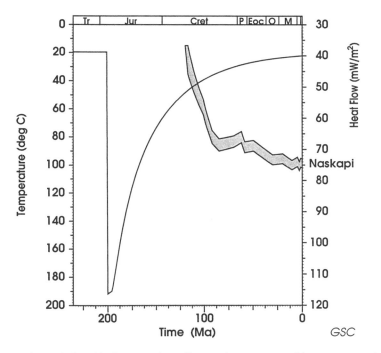

Figure 18. Relationship between heat flow and temperature history versus time of the Naskapi Member in North Triumph G-43.

On the other hand, the Naskapi source rock in Cohasset D-42 has a maturation of about 0.5% R_o and a high TOC content but contains Type IIB kerogen which has significantly higher activation energies. Therefore, hydrocarbon generation is extremely minor (Figure 13) and is not considered as a source rock for Cohasset D-42 light oil, although it shows some similarity with the light oils in oil-source rock correlation study. If the Naskapi Member at Cohasset also contained some Kerogen Type IIA-IIB, some amount of oil might have been generated, however, the oil would not have been expelled (Figure 14).

The upper Jurassic/Kimmeridgian source rock in Venture B-52 has a maturity of 1.0-1.45% R_o. Figure 16 shows the modeled maturation history, hydrocarbon generation, and migration through time of this source rock. It shows that the major hydrocarbon generation (both liquid and gas) occurred since 125 Ma with major expulsions of oil (or condensate) and gas lasting between 85 and 50 Ma. The cumulative amount of expelled liquids is about 50 barrels/acre ft. and gas is 90 barrels (equiv.)/acre ft.

Conclusions

* In modeling the interplay between the burial, compaction and thermal history of basin sediments we have derived estimates of the basins maturation framework which provides a basis for assessing the hydrocarbon generative capacity of identified source intervals.

* Our model-predicted thermal and maturity profiles are calibrated with reference to measured vitrinite reflectance and bottom hole temperatures. The higher maturity in the Cohasset D-42 well compared to Venture B-52 is caused by: (a) the difference in thermal conductivity in the lithostratigraphic profile (limestone, sandstone/shale package in Cohasset D-42 versus sandstone/shale in Venture B-52) and/or (b) the presence of older rocks at shallow depths in the Cohasset D-42; modeling, which takes rifting into account, suggests the older rocks were subjected to higher heat flow.

* The low activation energy profile and Arrhenius Constant of Kerogen Type IIA-IIB source rock suggests that 80% of the source rock can be converted to oil at a maturity around 0.8% R_o. This study and earlier studies (25) demonstrate the importance of using kerogen kinetic parameters specific to the source rocks encountered in the basin under study. The generation of liquid hydrocarbons can take place at low or higher levels of vitrinite reflectance depending on the kinetic parameters (activation energy profile and Arrhenius Constant) as suggested by Mukhopadhyay (27).

* The main expulsion of liquid hydrocarbons from the source rock of the Naskapi Member at N. Triumph G-43 has occurred since 25 Ma and about 20-30 barrels of oil/acre foot of source rock has been expelled. On the other hand, condensate and hydrocarbon gases, which were generated earlier, migrated from the upper Jurassic /Kimmeridgian source rocks in the Venture B-52 well between 80 and 50 Ma.

Acknowledgments

The authors acknowledge the assistance of Dr. Michael A. Kruge of Southern Illinois

University, Illinois, USA and Dan Jarvie of Humble Analytical Services, Humble, Texas for the analytical data and some interpretation on oil-source rock correlation (Dr. M. Kruge). The authors thank W. D. Smith, Bob Ryan, D. E. Buckley and China O. Leonard for reviewing the manuscript. The manuscript was handled by W. G. Dow for the review. This is Geological Survey of Canada Contribution Number 48693.

Literature Cited

1. Tissot, B. P.; Espitalie, J. *Rev. Inst. Fran. Petrol.* 1975, *30*, 743-777.
2. Welte, D. H.; Yukler, M. A. *Bull. Am. Assoc. Petrol. Geol.* 1981, *65*, 1387-1396.
3. Welte, D. H.; Yalcin, M. N. *Org. Geochem.* 1988, *13 (1-3)*, 141-151.
4. Sweeney, J. J.; Burnham, A. K. *Bull. Am. Assoc. Petrol. Geol.* 1989, *74*, 1559-1570.
5. Braun, R. L.; Burnham, A. K. *Energy & Fuels.* 1990, *4*, 132-146.
6. Ungerer, P.; et. al. *Bull. Am. Assoc. Petrol. Geol.* 1990, *74*, 309-334.
7. Lopatin, N. V. *Akad. Neuk S.S.S.R., Ser. Geol.* 1971, *3*, 95-106.
8. Waples, D. W. *Bull. Am. Assoc. Petrol. Geol.* 1980, *64*, 916-926.
9. Issler, D. R. *Can. Jour. Earth Sc.* 1984, *21*, 477-488.
10. Royden, L.; Keen, C. E. *Earth & Plan. Sc. Lett.* 1980, *51*, 343-361.
11. McCulloh, T. H.; Naeser, N. D. In *Thermal History of Sedimentary Basins: Methods and case Studies.* Naeser, N. D.; McCulloh, T. H., Eds.; Springer-Verlag: Berlin-Heidelberg, 1988, pp. 1-11.
12. Nantais, P. T. *Geol. Surv. Can. Open File Report 1175*, 1983, 179p.
13. Williamson, M. A.; DesRoches, K. *Bull. Can. Petrol. Geol.* 1993, *41*, 244-257.
14. Mudford, B.; Best, M. E. *Bull. Am. Assoc. Petrol. Geol.* 1989, *73*, 1383-1396.
15. Williamson, M. A.; Smyth, C. *Bull. Can. Petrol. Geol.* 1992. *40*, 136-150.
16. Mukhopadhyay, P. K.; Wade, J. A. *Bull. Can. Petrol. Geol.* 1990, *38*, 407-425.
17. Mukhopadhyay, P. K. *Geol. Surv. Can. Open File Report, 2620*, 1990, 49p.
18. Mukhopadhyay, P. K. *Geol. Surv. Can. Open File Report, 2921*. 1991, 99p.
19. Platte River Associates. In *BasinMod^{TM}: A Modular Basin Modelling System, Version 3.15*, Denver, Colorado, 1992.
20. Wade, J. A.; MacLean, B. C. In *Geology of the Continental Margins of Eastern Canada*, Keen, M. J.; Williams, L. J. Eds.; Geology of Canada, 1990, *2*, 167-238.
21. Palmer, A. R. *Geology.* 1983, *11*, 503-504.
22. Dehler, S. A. and Keen, C. E. *Can. Jour. Earth Sc.* 1993, *30*, 1782-1798.
23. Tissot, B. P; Welte, D. H. *Petroleum Formation and Occurrence*; Springer-Verlag: Berlin, Germany, 1984; 699p.
24. Keen, C. E.; Beaumont, C. In *Geology of the Continental Margin of Eastern Canada.* Keen, M. J.; Williams, G. L. Eds.; Geology of Canada, 1990, *2*, 391-472.
25. Hunt, J. M.; Lewan, M. D.; Hennet, R. J. C. *Bull. Am. Assoc. Petrol. Geol.* 1991, *75 (4)*, 795-807.
26. Mukhopadhyay, P. K. *Geol. Surv. Can. Open File 2804*, 1993
27. Mukhopadhyay, P. K. In *Diagenesis III: Developments in Sedimentology, 47*, Wolf, K. H.; Chillingarian, G. V., Eds.; 1992, 435-510.

Appendix 1

Modeling. We used the Platte River *BasinMod* program, which generates geohistory/ burial history using a geological model of stratigraphy versus time and utilizing a wide range of parameters such as lithology, absolute age, paleowater depth, surface temperature, compaction, permeability, conductivity, subsidence, thermal history using either rifting heat flow or present day bottom hole temperature or geothermal gradient. Compaction corrections have a significant impact on thermal history which, in turn, affects the timing of source rock maturity, petroleum generation and expulsion. For the modeling, the default compaction correction was used according to Sclater and Christie (19).

Our models assumed a Steady State Heat Flow which is based on the simple relationship between heat flow and thermal conductivity (Heat Flow = Thermal Conductivity X Temperature Gradient). Rifting heat flow was taken as the major heat source which usually decreases with time. One requirement for the calculation of rifting heat flow is a β factor, which is defined as β = Initial Lithospheric Thickness/Lithospheric Thickness Immediately After Stretching.

BMOD (19) provides a kinetic model, based upon the *Easy* %R_o model, developed at Lawrence Livermore National Laboratory (4) to calculate maturity in the form of %R_o. The model uses a distribution of Arrhenius rate constants to calculate global vitrinite maturation, which correlates maturation with reflectance.

The kinetic approach considers the diversity of composition and distribution of the original kerogen. It calculates multiple parallel reactions that occur as organic matter undergoes transformation to hydrocarbons. Each reaction has its own kinetic parameters such as percentage of kerogen with a specific activation energy and Arrhenius Constant or Frequency Factor (frequency with which certain reaction takes place). The kinetic approach considers the various HC potential for different kerogen. BMOD (19) converts kerogens to oil and gas (4-component model). The oil can undergo secondary cracking to gas and residue.

Expulsion efficiency reflects the amount of generated hydrocarbons that are expelled from the source rock according to percentages assigned to different maturity values (%R_o) for oil and gas. Almost all gas generated is expelled while 95% of generated oil is expelled. At a given vitrinite reflectance value, the expulsion efficiency is multiplied by the amount of hydrocarbon generation to get the amount expelled. The expulsion efficiency is defined as (original yield - final yield) / original yield. The efficiency of expulsion will vary with source rock richness and the nature of the pathway out of the source rock. Lowest expulsion is obtained for lean source rocks (<1% TOC) that are thick bedded (tens of meters). According to the Expulsion Efficiency VR method, the expulsion efficiency is correlated with vitrinite reflectance. The default values are 0.5% R_o, oil expulsion efficiency is 0.05 (5%), at 0.7% is 0.25 (25%), at 0.9% is 0.65 (65%), at 1.1% is 0.85 (85%) and at 1.3% is 0.95 (95%); 50% of gas is expelled around 0.7% R_o, 65% around 0.9%, 85% around 1.1%, 95% around 1.3%. Gas expulsion continued until 3.2% R_o.

The model requires input of lithology in relation to formation or absolute age, present day- and paleo-surface temperatures of each unit, water depth, present day bottom hole temperature or geothermal gradient, kerogen type and total organic carbon (TOC) content (in wt %), geochemical data such as Rock-Eval pyrolysis, vitrinite reflectance, etc.

RECEIVED June 3, 1994

Chapter 16

Apatite Fission Track Thermochronology Integrated with Vitrinite Reflectance

A Review

D. Arne and M. Zentilli

Fission Track Research Laboratory, Department of Earth Sciences, Dalhousie University, Halifax, Nova Scotia B3H 3J5, Canada

A review of basin studies for which both apatite fission track and vitrinite reflectance data are available suggests that total annealing of fission tracks in detrital apatite grains generally occurs at $R_o$max values in the range 0.7 to 0.9% for effective heating times between 10^6 and 10^8 years. This observation is in general agreement with predictions based on the comparison of kinetic models for annealing of fission tracks in apatite and the evolution of vitrinite reflectance. The integration of apatite fission track thermochronology and vitrinite reflectance is therefore not only favored by the complementary nature of the two techniques, it is also enhanced by the ability to predict a response in one system with data from the other.

Apatite fission track analysis is a thermochronological technique now extensively used for thermal history studies in sedimentary basins *(1-3)*. The technique exploits the inherent instability of latent fission tracks that form continuously through time from the spontaneous decay of trace ^{238}U in detrital apatite $[Ca_5(PO_4)_3(F,Cl,OH)]$ grains. Spontaneous fission tracks in apatite undergo significant annealing at temperatures between 20°C and 150°C over geologic time periods in the range 10^9 to 10^5 years, respectively. In the sedimentary basin environment, where effective heating times are generally on the order of 10^6 to 10^7 years, rigorous constraints on thermal history are only provided over the more restricted temperature range of ~60°C to 120°C. An evaluation of the density of etched spontaneous fission tracks on internal grain surfaces (Figure 1) and the determination of the distribution of etchable horizontal confined track lengths within the mineral grain allow an assessment of the timing, magnitude and style of cooling for rock samples that have been hotter in the past than they are at present day *(4, 5)*. This interpretive process

0097–6156/94/0570–0249$08.00/0

has been greatly facilitated by the development of predictive modelling techniques (6 - 10).

A progressive reduction in apatite fission track age with increasing down-hole temperature is demonstrated by data from detrital apatite grains in volcanogenic sandstone samples from the Otway Basin, Australia (Figure 2). The majority of apatite grains in these samples were deposited in the Early Cretaceous with fission track ages of ~120Ma and so the inheritance of fission tracks from older source areas can be ignored in this example (2). At down-hole temperatures greater than 90°C, individual apatite grains from each sample show a statistically significant spread in fission track age, with fission tracks in grains having the lowest chlorine contents being totally annealed at down-hole temperatures between 90°C and 95°C. Apatite grains with chlorine contents of ~2 wt. % are not totally annealed until ~125°C. This example clearly demonstrates the importance of apatite composition on the annealing behaviour of natural apatite grains in sedimentary basins.

A description of the physical processes involved in the formation and annealing of fission tracks in apatite, as well as in other materials, and a discussion of the various analytical methods in use are beyond the scope of the present review. The interested reader is referred to recent summaries for more information and an extensive list of references (11, 12). Of relevance to the present study is the comparison of kinetic descriptions for annealing of fission tracks in apatite and the evolution of vitrinite reflectance (R_o), as the integration of the two techniques is enhanced by the ability to use data from one technique to reliably predict a response in the other. Thus two published comparisons will be presented and evaluated using a compilation of studies in which both apatite fission track and vitrinite reflectance data are available. This evaluation will be followed by a series of conceptual examples illustrating the benefits of integrating the two thermal indicators and by a case study from the foothills region of southwest Alberta, Canada.

Evaluation of Kinetic Models

A major source of uncertainty in the application of apatite fission track and vitrinite reflectance data in basin studies lies in the selection of kinetic models used for predictive purposes. Some controversy exists as to whether it is appropriate to describe the annealing of fission tracks in apatite in terms of first-order kinetics (13), and more complex kinetic descriptions have been proposed (14, 15). Independent tests of the various annealing models in well-controlled geologic settings are available (16 - 18) but they are limited by the uncertainty inherent in comparing the annealing of fission tracks in detrital apatite grains having a range of chemical compositions with kinetic models derived from monocompositional apatite specimens. As previously demonstrated, the chlorine content of apatite is considered to be an important factor controlling the rate of track annealing at severe degrees of length reduction (2, 19), and a rigorous assessment of kinetic models in nature is not possible unless this variable is taken into account.

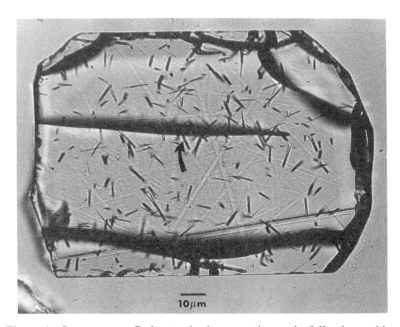

Figure 1. Spontaneous fission tracks in an apatite grain following etching with 5N HNO₃ for 20 seconds at room temperature. Scale bar is ~ 10 μm in length. Note the confined track at arrow intersecting the crack.

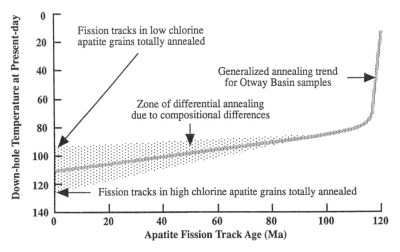

Figure 2. Decrease in apatite fission track age with increasing down-hole temperature, Otway Basin, Australia. Adapted from ref. 21.

Kinetic descriptions for the evolution of vitrinite reflectance are also not without controversy (20), with the result that predictive models are often adapted to suit the particular basin under investigation. The evaluation of kinetic models for vitrinite reflectance is further complicated by other factors such as maceral type, host rock lithology, sulphur content and analytical techniques (see other chapters in this volume). Given these uncertainties, we have not undertaken a detailed assessment of kinetic models, but rather have restricted our discussion to two published comparisons for annealing of fission tracks in apatite and the evolution of vitrinite reflectance.

A comparison between the evolution of vitrinite reflectance (derived from LOM values) and annealing of fission tracks in apatite for a variety of effective heating times and temperatures is presented in Figure 3 (1). The annealing data are from deep drill-holes and are based on the observed reduction in fission track age with increasing temperature. Of interest in this comparison is the prediction that contours describing the evolution of vitrinite reflectance and the annealing of fission tracks in apatite are oblique in time-temperature space. Thus the intersection of contours from the two separate thermal indicators will vary depending upon the effective heating time under consideration. For example, total annealing of fission tracks in apatite is predicted to correspond to an R_0 value of $\sim 0.65\%$ for an effective heating time of 10^6 years, while for an effective heating time of 10^9 years, total annealing of fission tracks in apatite is predicted to occur at an R_0 value of $\sim 1.0\%$ (Figure 3).

By contrast, a comparison of a model for annealing of fission tracks in Durango apatite having a Cl/F ~ 0.1 (21) with a distributed activation energy model for the evolution of vitrinite reflectance (22) is reproduced in Figure 4 (23). Fission track annealing depicted in this figure is based on the extrapolation of laboratory annealing experiments to geologic time and was determined using track length measurements rather than measured ages. Of note here is that, for the range of heating rates and temperatures shown, the contours describing the evolution of vitrinite reflectance and shortening of mean track length in apatite are parallel near the point where fission tracks are totally annealed in Durango apatite. The heating rates shown in Figure 4 correspond to effective heating times on the order of 10^5 to 10^7 yrs. Thus, the comparison of these two models predicts a constant R_0 value ($\sim 0.7\%$) for total annealing of fission tracks in apatite regardless of effective heating time. The precise R_0 value will vary somewhat depending on the composition of the apatite in question.

The two published comparisons described in the preceding paragraphs are in general agreement in that they predict total annealing of fission tracks in apatite at an R_0 value between $\sim 0.6\%$ and 0.8% for effective heating times of 10^6 to 10^8 years. Therefore, it should be possible to test this prediction using published thermal history studies for which both fission track and vitrinite reflectance data are available near the point at which total annealing of fission tracks has been inferred at some point in time. Unfortunately few specific studies have been reported in the literature, although it is possible to draw

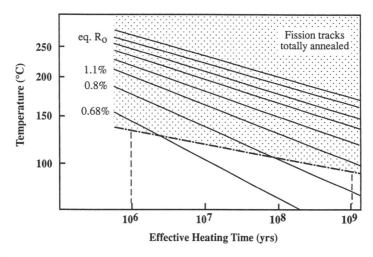

Figure 3. Relationship between annealing of fission tracks in apatite and the evolution of vitrinite reflectance. Note that the contours for the two systems are oblique in temperature - time space for the ranges given. Adapted with permission from reference 1. Copyright 1989 Springer-Verlag New York.

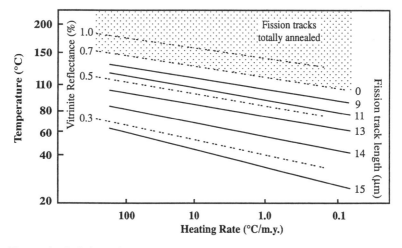

Figure 4. Relationship between annealing of fission tracks in apatite and the evolution of vitrinite reflectance. Note that the contours are parallel in time - temperature space near the point where fission tracks in apatite are totally annealed. Adapted from ref. 21.

some preliminary conclusions from regional data sets and by combining results from independent studies. Additional sources of error for a compilation such as this include differences in the composition of the apatite grains studied, the analytical procedures used to determine vitrinite reflectance, and the difficulty of estimating effective heating times, particularly where significant gaps in the stratigraphic record exist. Ideally, such comparative studies would be most useful in well sequences currently at maximum burial temperatures, but few studies of this nature are available in the literature.

The results of the compilation summarized in Table I indicate that total annealing of fission tracks in apatite generally occurs between $R_o$max values of 0.7% and 0.9% for a range of estimated effective heating times between $\sim 10^6$ years and 10^8 years. In addition, data from studies in which only an upper or lower bound on the point of total annealing are available are consistent with this observation. This range of between 0.7% to 0.9% $R_o$max for total annealing is slightly higher than that of 0.6% to 0.8% R_o predicted from the comparison of kinetic models discussed previously. Such a discrepancy is not surprising given the uncertainties inherent in such compilations from the literature and given the difficulty in determining the point of total annealing for fission tracks in samples that contain a range of apatite compositions, as fission tracks in different grains will totally anneal over a range of temperatures (2). Given the uncertainties in effective heating time for most studies, the comparisons presented in Figures 3 and 4 cannot be distinguished on the basis of the available data. Therefore it is not possible to determine whether total annealing of fission tracks in apatite of a particular composition occurs at a constant value of R_o regardless of effective heating time, or whether total annealing occurs over a range of R_o values depending on the effective heating time. Comparisons of R_o values with total annealing of fission tracks in apatite adjacent to small igneous intrusions, where the effective heating time is considerably less than 10^6 years, would help to resolve this uncertainty.

Examples of Integration

If it is accepted that fission tracks in apatite are totally annealed at $R_o$max values between 0.7% and 0.9% for effective heating times typical of sedimentary basins, then the integration of apatite fission track thermochronology with vitrinite reflectance data is enhanced by the ability to predict a response in one thermal indicator given data from the other. Integration can be discussed at two levels: where the two techniques complement each other and where one method is used to predict a response in the other. In terms of the two techniques complementing each other, the main areas to be considered are the timing of cooling, sample availability and temperature range covered by each technique, and the value of an independent check for each data set (e.g. 23, 24, 30, 32, 35). As a predictive tool, vitrinite reflectance data can serve as a guide to sampling drill core or surface rocks for apatite fission track thermochronology by identifying those samples in which fission tracks have been totally annealed prior to a phase of cooling.

Table I: Correlation of R_o values with total annealing of fission tracks in apatite

Study area (well name)	R_o value for total annealing	Estimated effective heating time	Reference	Comments
1) Total annealing of fission tracks bracketed by vitrinite reflectance data				
Taranaki Basin, New Zealand (1 Fresne)	0.7 - 0.8%	~1 m.y.	(24)	No details given
South Island, New Zealand	~0.7%	1-10 m.y.	(25)	No details given
ANWR, Northern Alaska	0.6 - 0.8%	1-10 m.y.	(26)	Fluorapatite; no R_o details given
Amery Group, East Antarctica	0.7 - 0.9%	1-100 m.y.	(27)	Dominantly fluorapatite; R_o max
Canning Basin, Australia (Grevillea 1)	~0.8%	1-100 m.y.	(28, 29)	Dominantly fluorapatite; R_o eq
2) Upper or lower limit on total annealing of fission tracks in apatite				
Southern Israel	<0.76- 0.95%	1-100 m.y.	(30)	R_o max[a]
New York state, U.S.A.	>0.7%	1-100 m.y.	(31)	R_o max
Midlands Shelf, U.K. (47/29a-1)	>0.7%	1-100 m.y.	(32)	R_o max
Maritimes Basin, Canada (L-49 and E-95)	<0.8-0.9%	1-100 m.y.	(33)	R_o max

[a] Converted from R_o random using the relationship $\%R_o$ max $= (\%R_o \text{random} - 0.0012)/0.94$ (34)

Integration of the two techniques is also useful where the effects of rapid heating of a sediment shortly after deposition cannot be resolved on the basis of fission track data alone.

Complementary Advantages. Aside from exceptional cases where R_o profiles are offset across unconformities *(36, 37)*, vitrinite reflectance data provide little constraint on the absolute time of heating where it can be demonstrated that a stratigraphic sequence has been hotter in the past than at present-day. Generally only an upper limit on the time of heating can be obtained, based on the stratigraphic age of the youngest unit in a well or region to show R_o values higher than predicted for the present-day thermal regime. The question of relative timing also becomes important for the evaluation of hydrocarbon charge models where it is desirable to determine whether maximum temperatures developed prior to or after the formation of potential structural traps *(38)*.

The absolute time of cooling from maximum temperatures can generally be inferred from apatite fission track data provided that temperatures for the samples have not greatly exceeded the point at which fission tracks are totally annealed in apatite (i.e., $0.7\% < R_o < 0.9\%$), in which case only broad constraints on the time of cooling are available if cooling was gradual (Figure 5; path B). Fission tracks in a sample following path A in Figure 5 were just totally annealed following deposition, allowing an estimate of T_1, the time of cooling from maximum paleotemperatures, based on predictive modelling of fission track data from the sample *(e.g., 6, 9)*. Fission tracks in a sample following path B were also totally annealed following deposition, but track retention did not commence until T_2. Thus the time of cooling from maximum paleotemperatures is only constrained to lie somewhere between D_1, the depositional age, and T_2. In addition, the maximum paleotemperatures to which the samples were exposed are only broadly constrained by apatite fission track data (i.e., $> \sim 110^\circ C$), although these may be inferred from vitrinite reflectance data if the effective heating time can be estimated.

If cooling was rapid, the time of cooling from maximum paleotemperatures can generally be constrained regardless of the magnitude of cooling, as shown by paths A' and B' in Figure 5. Note that the time of cooling may be directly inferred from the apatite fission track age of a sample following path A' provided the sample has cooled to temperatures at which fission tracks are relatively stable (i.e., below $\sim 60^\circ C$). The time of cooling can also be inferred for sample B', but this would first require taking into account the effects of annealing at the present-day temperature of the sample *(e.g., 24)*. Thus a prior review of available vitrinite reflectance data can serve to predict whether apatite fission track data will provide rigorous constraints on the time of cooling from maximum temperatures, although ultimately this may depend upon the rate of cooling.

The lithologies to be sampled for an apatite fission track or a vitrinite reflectance study are generally different. While the former utilizes coarse clastic rocks, or even crystalline basement samples, the latter generally targets fine clastic units and coal seams. Thus sampling for both techniques can provide a more even distribution of samples through the stratigraphic section under investigation. Use of apatite fission track data also provides rigorous paleotemperature control in rocks of pre-Devonian age that do not contain vitrinite. In addition, while oxidation is a concern when sampling for vitrinite reflectance at the surface, the effect of weathering on fission tracks in apatite is generally negligible, except in extreme cases *(39)*.

A major limitation on the use of apatite fission track data as a paleotemperature indicator results from total annealing of fission tracks in apatite at temperatures of around $110^{\circ}C$ for effective heating times on the order of 10^7 years *(21)*, the precise value depending on the composition of the apatite grains in question and the kinetic description employed *(18)*. Use of vitrinite reflectance data is therefore required to extend the temperature range under investigation beyond $\sim 110^{\circ}C$. This is particularly desirable where one of the objectives of the thermal history study is to constrain the paleotemperature gradient. Shown in Figure 6 is the range of paleotemperature gradients in a well that are broadly compatible with paleotemperature estimates derived from apatite fission track data alone (light shading), and those derived using additional estimates of maximum paleotemperature from vitrinite reflectance data at greater depths in the well (dark shading). Note that the use of both apatite fission track and R_o data to constrain paleotemperatures within a single vertical section would be facilitated by a similarity in their kinetic response *(23)*.

In general, each paleotemperature estimate will have an associated uncertainty depending on which kinetic model is used, the thermal history undergone by the samples, modelling procedures used to derive the paleotemperature estimates, compositional variation, and uncertainties in the effective heating time. In addition, the inheritance of fission tracks formed prior to the deposition of detrital apatite grains may also be important. In practice, it is difficult to quantify all possible source of uncertainty in deriving paleotemperature estimates for individual samples in a vertical sequence. For the purposes of illustration in Figure 6, nominal error bars of $\pm 10^{\circ}C$ have been used. The range of paleotemperature gradients shown for a given data set are those that are consistent with the error bars for each paleotemperature estimate.

As in the case of any analytical technique, it is useful to have an independent check on the validity of the data being obtained. In some instances, such as in the case of clastic rocks containing detrital apatite grains and dispersed organic matter, it is possible to derive paleotemperature estimates from a single sample using the two different techniques. Using the model for annealing of fission tracks in Durango apatite *(21)* and the distributed activation energy model for the evolution of vitrinite reflectance *(22)*, similar paleotemperature estimates are predicted from the same sample for any given thermal history *(23)*. The availability of an independent check is perhaps of more value for assessing R_o data, particularly in areas where

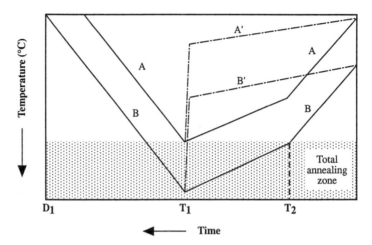

Figure 5. The influence of the magnitude and rate of cooling on constraining the time of cooling from maximum paleotemperatures.

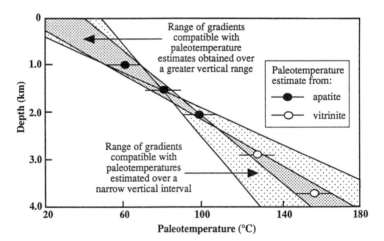

Figure 6. The influence of sampling range on constraining paleotemperature gradients. Error bars are nominal at $\pm 10^{\circ}C$. The light stippled area is constrained by paleotemperature estimates from apatite alone, whereas the darker area includes additional constraints from vitrinite reflectance data.

suppression of vitrinite reflectance or contamination by re-worked vitrinite is suspected.

Predictive Advantages. As previously demonstrated in Figure 5, the best constraint on the time of cooling available from apatite fission track data is provided by samples in which fission tracks in apatite were totally annealed prior to rapid cooling as these samples are interpreted to have only retained fission tracks since the time of cooling. The establishment of an equilibrium annealing profile in a vertical sequence (see the generalized trend in Figure 2) followed by rapid cooling produces a characteristic "break-in-slope" in the fission track age profile *(1, 40)*. Samples directly below the "break-in-slope" provide the best constraint on the time of cooling and their identification in well sequences and vertical outcrop suites forms an important aspect of fission track sampling strategies. If it is accepted that fission tracks in most apatite samples are totally annealed at $R_o$max values in the range 0.7% to 0.9%, vitrinite reflectance data can be used to direct sampling toward those samples likely to give the most information regarding the time of cooling (Figure 7).

Following similar reasoning, vitrinite reflectance data can be used to evaluate simple structures at high levels in fold and thrust belts across which it may be possible to constrain the time of movement. Provided maximum paleotemperatures in an upthrown block predate the time of faulting, a break in thermal maturity should occur across the fault exposed at the surface. This situation is represented in Figure 8 by an offset in maximum paleotemperature isotherms, lines of constant maximum paleotemperature, similar to the stratigraphic offset across the fault. If maximum paleotemperatures post-date faulting (a situation favorable for the accumulation of hydrocarbons in structural traps), then no offset in maximum paleotemperature isotherms would be expected and no break in thermal maturity will occur across the fault at surface. For the situation illustrated in Figure 8, vitrinite reflectance data can be used to establish that a break in thermal maturity exists across a structure.

If cooling of the upthrown block accompanied faulting, then the time of cooling inferred from apatite fission track data will give the time of thrusting. In such a situation it is desirable to sample an upthrown block in which fission tracks in apatite were totally annealed immediately prior to faulting, as these samples provide the best constraint on the time of cooling. Whether a break in fission track age is preserved across the fault depends on the amount and time of cooling in the downthrown block, as the present-day erosional surface may represent a deeper structural level that was exhumed following or during thrusting. Where an upthrown block has been juxtaposed against a relatively cool downthrown block in which fission tracks have only been partially annealed, a fission track age discontinuity is likely to be preserved. Vitrinite reflectance data can be used to identify such favorable situations (Figure 8), and in addition, also provide information on the timing of structural development relative to the timing of maximum paleotemperatures *(38)*.

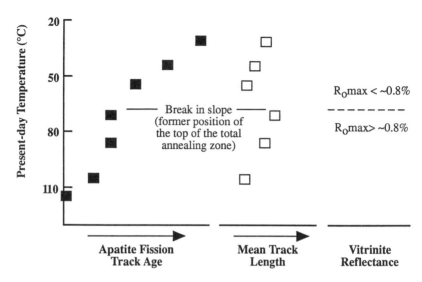

Figure 7. Use of vitrinite reflectance data in a well sequence to predict the interval below which fission tracks in apatite were totally annealed prior to an episode of cooling.

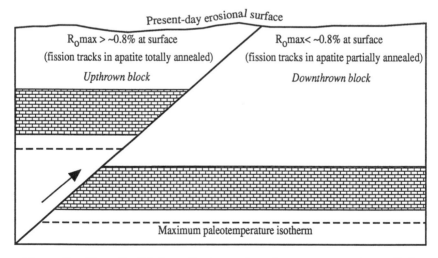

Figure 8. Use of vitrinite reflectance data from across a thrust fault exposed at the surface to predict the presence of a discontinuity in fission track age data that can be used to estimate the absolute time of thrusting.

In some tectonic settings sediments may be heated immediately after deposition and then cool very quickly (Figure 9). In these circumstances, the measured apatite fission track age of a sample may overlap the depositional age when analytical and/or stratigraphic uncertainties are taken into account. Thus, on the basis of the fission track data alone, it may not be possible to determine whether the apatite grains were cooled immediately before or after deposition. Vitrinite reflectance data can be used to predict whether post-depositional heating took place, provided re-working of vitrinite can be discounted, and thus allow the interpretation of fission track data that would otherwise be equivocal.

For example, apatite in a sedimentary rock sample with a fission track age that overlaps its stratigraphic age might be derived from a rapidly eroding source area nearby or might be volcanogenic in origin. However, if vitrinite from nearby coal seams give an $R_o$max value greater than 0.8%, then it can be inferred that the sequence had been heated to temperatures sufficient to totally anneal fission tracks in apatite, but cooled shortly after deposition to give an apatite fission track age indistinguishable from its stratigraphic age.

Burnt Timber Thrust, Southwest Alberta, Canada

Regional Setting. The timing of thrust movement in the Foothills region of southwest Alberta can be broadly constrained by the stratigraphic ages of sediments deformed near individual structures and is generally considered to decrease in age from west to east as the axis of the Alberta Basin is approached *(41)*. However, the absolute timing of final thrusting is not independently known and is only constrained to have been younger than the early Paleocene sedimentary rocks disrupted by tectonism. Apatite fission track thermochronology has previously been used to successfully constrain the absolute time of movement across individual faults at high structural levels *(42-44)*. Although the interpretation of thermochronological data proximal to ductile structures is complicated by the possible effects of shear heating *(45,46)*, vitrinite reflectance data near major structures in the Front Ranges of southwest Alberta indicate that shear heating was generally not a significant factor at the structural level presently exposed *(47)*.

A simplified geological map of southwest Alberta is provided in Figure 10 *(48)*. An upper limit on the time of movement across the Burnt Timber Thrust is provided by the stratigraphic age of the youngest deformed unit exposed in the axis of the Williams Creek Syncline. This unit is inferred to be lower Paleocene in age on the basis of palynological studies (A. Sweet, Geological Survey of Canada, personal communication, 1993), which constrain the time of thrusting to be post-65 Ma to 60 Ma in age *(49)*. Lower Paleocene rocks in the footwall of the thrust are in contact with Upper Jurassic to Lower Cretaceous rocks of the Kootenay Group, indicating a stratigraphic offset of at least 3 km, measured from the base of the Paleocene *(50)*. Vitrinite reflectance and apatite fission track data were obtained from the same surface

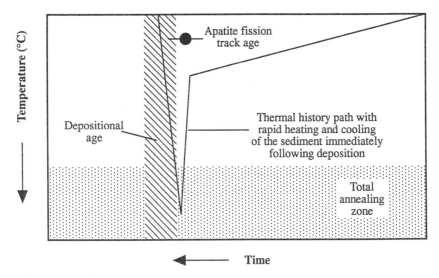

Figure 9. Use of vitrinite reflectance and apatite fission track data to indicate rapid heating and cooling of a sediment immediately after deposition.

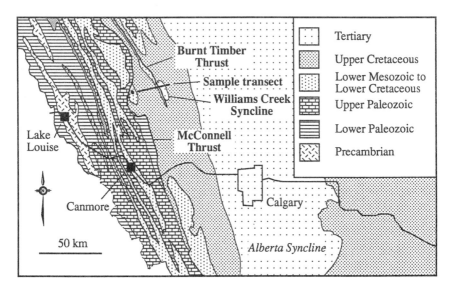

Figure 10. Geological map of southwest Alberta showing the location of the transect across the Burnt Timber Thrust (heavy line). Adapted with permission from ref. 48. Copyright 1975 Canadian Society of Petroleum Geologists, Calgary.

rock samples in a transect perpendicular to the trace of the Burnt Timber Thrust along the Red Deer River.

Interpretation of Thermal History. Vitrinite reflectance data from a profile across the Burnt Timber Thrust are shown in Figure 11. There is a general increase in mean R_0max values from 0.89% to 1.54% with increasing stratigraphic age in the hanging wall of the Burnt Timber Thrust that is consistent with maximum paleotemperatures having been attained by this section during burial prior to thrusting. By contrast, mean R_0max values in the core of the Williams Creek Syncline vary from 0.61% to 0.55%. Based on the relationship between total annealing of fission tracks in apatite and R_0max established previously in this review, these data suggest that fission tracks in apatite from surface samples in the footwall of the Burnt Timber Thrust were not totally annealed at any time following deposition. By contrast, fission tracks in apatite from the hanging wall samples were totally annealed at some time following deposition and thus data from these samples can be interpreted in terms of the time at which track retention began. On the basis of these vitrinite reflectance data, this structure was selected for detailed apatite fission track thermochronology.

A significant break in thermal history across the Burnt Timber Thrust is also indicated by preliminary apatite fission track data from across the structure (Figure 12). The two samples furthest to the west along the transect give pooled fission track ages of 63 ± 11 and 64 ± 8 Ma, respectively, (all uncertainties are quoted at 2 sigma) that are significantly younger than their late Cretaceous stratigraphic ages. This observation, coupled with the fact that both samples give mean lengths near 14 μm, suggests that fission tracks were totally annealed in these samples after deposition, followed by rapid cooling during the early Tertiary at close to 64 ± 6 Ma (weighted mean value). A sample of Upper Jurassic to Lower Cretaceous Kootenay Group quartzite and a sample of Lower Cretaceous Blairmore Group sandstone from the direct hanging wall of the Burnt Timber Thrust give identical apatite fission track ages of 59 ± 7 Ma, with mean track lengths of $14.0\pm0.4\mu$m and $14.3\pm0.4\mu$m, respectively. Thus these samples are interpreted to have cooled rapidly during the early Tertiary, at 59 ± 5 Ma (weighted mean value). Two samples of early Paleocene sediment from the direct footwall of the Burnt Timber Thrust give apatite fission track ages of 76 ± 9 Ma and 82 ± 15 Ma, respectively, that are significantly older than the inferred time of hanging wall cooling, as well as being older than the stratigraphic age of the sediment (i.e., between 65 and 60 Ma).

A sample of Upper Cretaceous Brazeau Formation from the proximal footwall of the Burnt Timber Thrust also gives a relatively high R_0max value indicative of total annealing of fission tracks in apatite. The fission track age of this sample is intermediate to those of samples from either side on the fault, suggesting either some thermal complexity immediately adjacent to the fault or unusual annealing behaviour. Microprobe analysis of this sample indicates that it contains a dominant population of apatite grains containing an unusually

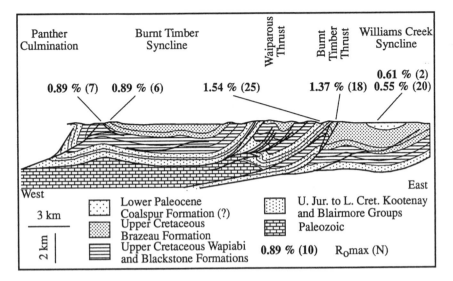

Figure 11. Geological cross section through the Burnt Timber Thrust showing R_0max data. Cross section adapted with permission from ref. 50. Copyright 1978 Geological Survey of Canada, Ottawa. Data available from the authors on request.

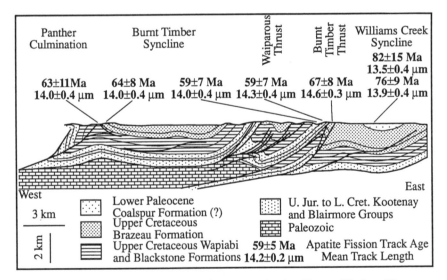

Figure 12. Geological cross section through the Burnt Timber Thrust showing apatite fission track data. Uncertainties quoted at ±2 sigma. Cross section adapted with permission from ref. 50. Copyright 1978 Geological Survey of Canada, Ottawa. Data available from the authors on request.

high amount of chlorine (mean of 1.75 wt. % Cl for 14 grains) compared to apatite grains from the Kootenay and Blairmore Groups (typically less than 1.0 wt. % Cl). Apatite grains in this sample are therefore inferred to have begun track retention prior to the Kootenay and Blairmore Group samples because they are of a relatively more track retentive composition *(e.g., 19)*.

Apatite fission track data from the hanging wall of the Burnt Timber Thrust can be used to constrain the time of movement across the fault if it assumed that denudation of the hanging wall kept pace with the rate of thrusting. Alternatively, the data could also be interpreted to reflect a dramatic contrast in heat flow or thermal conductivity caused by the juxtaposition of different rock types across the fault, with final cooling of both the hanging wall and footwall coeval with regional exhumation. Such an explanation is considered to be unlikely given the large difference in maximum temperatures inferred for surface samples of similar lithology from across the structure. Significant differences in surface elevation across the Burnt Timber Thrust prior to regional exhumation can also be precluded as a possible explanation for the observed contrast in thermal history given the short horizontal distance between samples. Thus, as a preliminary estimate, movement across the Burnt Timber Thrust is inferred to have occurred at 59 ± 5 Ma, shortly after deposition of lower Paleocene sediments preserved in the Williams Creek Syncline.

Conclusions

A compilation of thermal history studies in which both apatite fission track and vitrinite reflectance data are available suggests that total annealing of fission tracks in apatite occurs at R_0max values in the range 0.7% to 0.9% for effective heating times between 10^6 and 10^8 years. Two published comparisons of kinetic models for annealing of fission tracks in apatite and the evolution of vitrinite reflectance *(1, 23)* cannot be distinguished on the basis of the available data without comparisons involving effective heating times less than 10^6 years. An integrated approach to basin thermal history and structural studies using apatite fission track thermochronology and vitrinite reflectance is therefore not only favored by the complementary nature of information provided by the two techniques, but is also enhanced by the ability to predict a response in one system using data from the other.

Acknowledgments

A review article of this nature must necessarily draw extensively upon the work of others and would not be possible without the contribution of those authors cited. In particular, DA would like to acknowledge his former employers at Geotrack International, Australia for introducing him to the benefits of integrating apatite fission track thermochronology and vitrinite reflectance data. Vitrinite reflectance data and maceral descriptions for the Burnt Timber area were provided by Alan Cook and associates at Keiraville Konsultants,

Australia. Financial support for the Dalhousie University Fission Track Research Laboratory has been provided by grants from the Natural Sciences and Engineering Research Council of Canada (NSERC), grants from Amoco Canada, Husky Oil, LASMO Nova Scotia and Mobil Canada under the Industry Research Associates Scheme, and from Imperial Oil Resources. DA would also like to acknowledge the support of a Dalhousie University Killam Fellowship during the preparation of this review. This manuscript benefited considerably from a critical review by Nancy Naeser, U.S. Geological Survey.

Literature Cited

1.   Naeser, N.D.; Naeser, C.W; McCulloh, T.H. In *Thermal History of Sedimentary Basins - Methods and Case Histories;* Naeser, N.D.; McCulloh, T.H., Eds.; Springer-Verlag: New York, New York, **1989**; pp 157-180.
2.   Green, P.F.; Duddy, I.R.; Gleadow, A.J.W.; Lovering, J.F. In *Thermal History of Sedimentary Basins - Methods and Case Histories*; Naeser, N.D.; McCulloh, T.H., Eds.; Springer-Verlag: New York, New York, **1989**; pp 181-195.
3.   Naeser, N.D. In *Basin Modelling: Advances and Applications;* A.G. Doré et al., Eds.; Elsevier: Amsterdam, 1993; pp 147-160.
4.   Crowley, K. *Nuclear Tracks* **1985**, 10, 311-322.
5.   Gleadow, A.J.W.; Duddy, I.R.; Green, P.F.; Lovering, J.F. *Contr. Mineral. Pet.* **1986**, 94, 405-415.
6.   Green, P.F.; Duddy, I.R.; Gleadow, A.J.W.; Hegarty, K.A.; Laslett, G.M.; Lovering, J.F. *Chem. Geol. (Isot. Geosci. Sect.)* **1989**, 79, 155-182.
7.   Lutz, T.M.; Omar, G. *Earth Planet. Sci. Lett.* **1991**, 104, 181-195.
8.   Corrigan, J. *J. Geophys. Res.* **1991**, 96, 10,347-10,360.
9.   Willet, S.D. In *Short Course Handbook on Low Temperature Thermochronology;* Zentilli, M.; Reynolds, P.H., Eds.; Mineralogical Association of Canada: Nepean, Ontario, 1992; pp 43-71.
10.   Crowley, K.D. *Comp. Geosci.* **1993**, 19, 619-626.
11.   Wagner, G.; Van den Haute, P. *Fission-Track Dating;* Kluwer Academic Publishing: Hingham, Massachusetts, 1992.
12.   Ravenhurst, C. E.; Donelick, R. A. In *Short Course Handbook on Low Temperature Thermochronology;* Zentilli, M.; Reynolds, P.H., Eds.; Mineralogical Association of Canada: Nepean, Ontario, 1992; pp 21-42.
13.   Green, P.F.; Duddy, I.R.; Laslett, G.M. *Earth Planet. Sci. Lett.* **1988**, 87, 216-228.
14.   Carlson, W.D. *Am. Mineral.* **1990**, 75, 1120-1139.
15.   Crowley, K.D.; Cameron, M.; Schaefer, R.L. *Geochim. Cosmochim.* **1991**, 55, 1449-1465.
16.   Gleadow, A.J.W.; Duddy, I.R. *Nuclear Tracks* **1981**, 5, 169-174.
17.   Vrolijk, P.; Donelick, R. A.; Queng, J.; Cloos, M. *Proc. ODP, Sci. Res.* **1992**, 129, 169-176.
18.   Corrigan, J.D. *Chem. Geol. (Isot. Geosci. Sect.)* **1993**, 104, 227-249.

19. Green, P.F.;Duddy, I.R.;Gleadow, A.J.W.;Tingate, P.R.;Laslett, G.M. *Chem. Geol. (Isot. Geosci. Sect.)* **1986**, 59, 237-253.
20. Morrow, D. W.; Issler, D. R. *Am. Assoc. Pet. Geol. Bull.* **1993**, 77, 610-624.
21. Laslett, G.M.; Green, P.F.;Duddy, I.R.; Gleadow, A.J.W. *Chem. Geol. (Isot. Geosci. Sect.)* **1987**, 65, 1-13.
22. Burnham, A.K.;Sweeney, J.J.*Geochim. Cosmochim.* **1989**,53,2649-2657.
23. Duddy, I.R.; Green, P.F.; Hegarty, K.A.; Bray, R. *Offshore Australia Conference Proceedings*; Melbourne, Victoria, 1991; Vol. 1, pp 49-61.
24. Kamp, P.J.J.;Green, P.F.*Am. Assoc. Petrol. Geol. Bull.*, **1990**, 74, 1401-1419.
25. Seward, D. *Chem. Geol. (Isot. Geosci. Sect.)* **1989**, 79, 31-48.
26. O'Sullivan, P.B.; Green, P.F.; Bergman, S.C.; Decker, J.; Duddy, I.R.; Turner, D.L. *Am. Assoc. Petrol. Geol. Bull.*, **1993**, 77, 359-385.
27. Arne, D.C. *Ant. Sci.* in press.
28. Arne, D.C.; Green, P.F.; Duddy, I.R.; Gleadow, A.J.W.; Lambert, I.B.; Lovering, J.F. *Aust. J. Earth Sci.* **1989**, 36, 495-513.
29. Ellyard, E.J. In *The Canning Basin;* Purcell, P.G., Ed.; Proceedings of the Geological Society of Australia/Petroleum Exploration Society of Australia: 1984; pp 359-375.
30. Feinstein, S.; Kohn, B.P.; Eyal, M. In *Thermal History of Sedimentary Basins - Methods and Case Histories;* Naeser, N.D.; McCulloh, T.H., Eds.; Springer-Verlag: New York, New York, 1989; pp 197-217.
31. Miller, D.S.; Duddy, I.R. *Earth Planet. Sci. Lett.* **1989**, 93, 35-49.
32. Bray, R.; Green, P.F.; Duddy, I.R. In *Exploration Britain: Into the Next Decade;* Hardman, R.F.P. Ed.; Geological Society of London Special Publication No. 67, 1992; pp 3-25.
33. Ryan, R.J.; Grist, A.M.; Zentilli, M. In *Short Course Handbook on Low Temperature Thermochronology;* Zentilli, M.; Reynolds, P.H., Eds.; Mineralogical Association of Canada: Nepean, Ontario, 1992; pp 141-155.
34. England, T.D.J.; Bustin, R.M. *Bull. Can. Pet. Geol.* **1986**, 71-90.
35. Issler, D.R.; Beaumont, C.; Willet, S.D.; Donelick, R.A.; Mooers, J.; Grist, A. *Bull. Can. Petrol. Geol.* **1990**, 38A, 250-269.
36. Dow, W.G. *J. Geochem. Expl.* **1977**, 7, 79-99.
37. Katz, B.J.; Pheifer, R.N.; Schunk, D.J. *Am. Assoc. Petrol. Geol. Bull.* **1988**, 72, 926-931.
38. Underwood, M.B.; Fulton, D.A.; McDonald, K.W. *J. Petrol. Geol.* **1988**, 11, 325-340.
39. Gleadow, A.J.W.; Lovering, J.F. *Earth Plant. Sci. Lett.* **1974**, 22, 163-168.
40. Gleadow, A.J.W.; Fitzgerald, P.G. *Earth Planet. Sci. Lett.* **1987**, 82, 1-14.
41. McMechan, M.E.; Thompson, R.I. In *Western Canada Sedimentary Basin: a Case Study;* Ricketts, R.D., Ed.; Canadian Society of Petroleum Geologists: Calgary, Alberta, 1989, pp 47-72.

42.   Wagner, G.A.; Michalski, I.; Zaun, P.; *The German Continental Deep Drilling Program (KTB);* Springer-Verlag: Heidelberg, 1989, pp 481-500.
43.   Hill, K.C.; Gleadow, A.J.W. *Aust. J. Earth Sci.* **1989**, 39, 515-539.
44.   Fitzgerald, P.G. *Tectonics* **1992**, 11, 634-662.
45.   Kamp, P.J.J.; Green, P.F.; White, S. *Tectonics* **1989**, 8, 169-195.
46.   Tagami, T.; Lal, N.; Sorkhabi, R.B.; Nishimura, S. *J. Geophys. Res.* **1988**, 93, 13,705-13,713.
47.   Bustin, R.M. *Tectonophysics* **1983**, 95, 309-328.
48.   Jackson, P.C. Geological Highway Map of Alberta, Canadian Society of Petroleum Geologists, **1975**.
49.   Harland, W.B.; Armstrong, R.L.; Cox, A.V.; Craig, L.E.; Smith, A.G.; Smith, D.G. *A Geologic Time Scale;* Cambridge University Press: Cambridge, 1989.
50.   Ollerenshaw, N.C., Geological Survey of Canada Map 1457A, **1978**.

RECEIVED April 15, 1994

Chapter 17

Paleoheat Flux Reconstruction from Thermal Indicators

He Wei[1], Malvin Bjorøy[2], and Elen Roaldset[1]

[1]Department of Geology and Mineral Resources Engineering,
Norwegian Institute of Technology, 7034 Trondheim, Norway
[2]Geolab Nos A/S, P.O. Box 5740 Fossegrenda, 7002 Trondheim, Norway

A numerical method is presented for reconstructing palaeoheat flux from thermal maturity indicators. This method utilises kinetic models or empirical expressions of thermal indicators and allows specification of various dependencies of heat flux on time such as exponential, constant, linear, parabolic, polynomial, and even a free variation with geological age, i.e. the variation of heat flux does not follow any time-dependent function. The chemistry-based kinetic models of vitrinite reflectance are valid in controlling palaeoheat flux at a maturity of >0.4 %R_o. The Heat flux of geologically younger sediments, which makes a large contribution to maturity, may be more accurately reconstructed by using a parallel reaction model of smectite/illite conversion kinetics. Of the kinetic models of thermal indicators, it is considered better to use an activation energy distribution model rather than a single reaction model, since the kinetic parameters used in parallel reactions are less sensitive, besides that the parallel reaction model describes the formation and conversion of thermal indicators under geological conditions more completely.

Palaeoheat flux and temperature play a very important role in subsurface organic and mineralogical reactions, such as kerogen degradation and the transformation of biomarkers and of smectite to illite. Much attention has therefore been paid to the reconstruction of palaeo-thermal histories of sedimentary basins (1-5). Palaeoheat flux can be estimated by either geophysical or geochemical methods. Geophysical methods, such as the tectonic subsidence approach (1) in rift basins, only provide for an exponential variation of heat flow with geological time. Evaluation of thermal history based on maturity indicators is now an accepted practice since these indicators record the time-temperature events of sediments. The thermal indicators will be widely accepted in coming years due to their specificity of temperature and the increasing ease of obtaining this type of data (2). Modern basin modelling should not use the

0097-6156/94/0570-0269$08.00/0
© 1994 American Chemical Society

inaccurate heating parameters such as temperature gradient with depth (°C/km) or heating rate (°C/ma).

A "thermal indicator" is defined here as an indicator whose value changes only with time-temperature events, examples of such indicators are the isomerisation and aromatisation of biomarkers, increase in vitrinite reflectance, and the smectite to illite conversion.

Basement heat flow is the main thermal energy source which determines formation temperature. "Present-day heat flux", the heat currently flowing from the interior of the earth toward the surface, can be calculated fairly accurately from down-hole temperature measurements. However, "palaeoheat flux", the heat which has flowed through sediments in the past, can only be assessed using models.

Inverse modelling of thermal indicators to estimate palaeoheat flux has been used by some groups (ex. *3-5*). In order to find a more efficient and sophisticated heat flux reconstruction method using the thermal indicators, we performed numerical experiments on classical optimisation algorithms. The results show that, by introducing a minimisation procedure into the palaeoheat flux reconstruction, our method allows specification of various dependencies of heat flux on geological time and even a free variation of heat flux with time. The method and its uncertainties are discussed in detail below and two examples of applications are shown.

Method

The method assumes constant heat flow with depth, i.e. that the heat flow changes only with geological time. This assumption is a condition of any one-dimensional steady-state heat conduction model. By definition, the alteration and transformation of the thermal indicator (TID) is a function of time-temperature events, and its value directly corresponds to the time-temperature integral (TTI):

$$TID = F \left(-A \int exp(-E/RT(t,z)) \ dt \right)$$

$$= F (\ TTI \ ) \qquad\qquad [1]$$

In this expression, E and A are activation energy and frequency factor respectively for the transformation of the thermal indicator's molecules. The value of TID is usually a function of a transformation ratio that is determined by TTI (ex. *6-8*).

This method is a one-dimensional heat flux reconstruction based on observed values of the thermal indicator at present-day depths, as illustrated by the diagram of the burial history of one well (Figure 1). The observed TID's values are indicated by 1, 2, 3, ... in the diagram. The corresponding burial lines were calculated by a decompaction model. Where Q0 is present-dat heat flux, Q1, Q2, Q3, Q4 are the unknown palaeoheat fluxes at corresponding ages and $\Delta TTI0$, $\Delta TTI1$, $\Delta TTI2$, $\Delta TTI3$, $\Delta TTI4$ are the differences of the time-temperature integrals between the corresponding ages.

For a given present-day depth (Figure 1), the geological time duration of equation [1] is fixed. The temperature history for this point, T(t,z) which is a function of time (t) and depth (z), is determined by heat flow history and the thermal

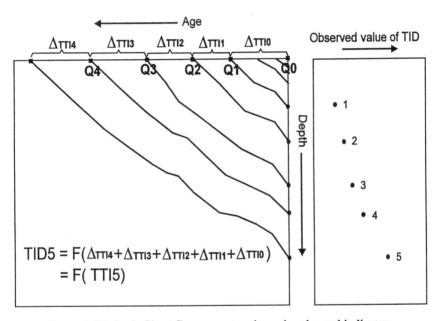

Figure 1. Method of heat flux reconstruction using thermal indicators.

conductivity of the sediments; and both vary with geological age. If it is possible to estimate the thermal conductivity based on the decompaction model, then by adjusting heat flow values along the burial time, it is possible to let the computed TID value close toward that observed at this point. Fitting single points is usually easy, however the solution is not unique. Two or more different sets of palaeoheat flux values may result in the same computed TID value (see uniqueness demonstration below). For this reason, more than one observed TID value is required to restrict the palaeoheat flux.

In the case of several observed values of the TID at one well, adjusting heat flow history to approach one point may result in a poor fit of the other points since the temperature histories of all points are determined by the same heat flow history under the assumption of constant heat flow with depth. A best fit for all the points is the intention of this method. The method utilises optimisation algorithms to search for an appropriate set of palaeoheat flux values that minimise the total deviation of computed TID values from that observed at various present-day depths.

The method described aims to determine palaeoheat flux Q(t), which varies with time (t), where the objective function of

$$\sum_{j=1}^{m} (\, \text{Mtid}[j] - \text{Ctid}[j]\,)^2 \qquad [2]$$

takes on a minimum value. This is an optimisation problem where m denotes the number of sampling points. Mtid[j] is the observed value of the thermal indicator at a present-day depth point j, and Ctid[j] is the computed value at the same depth using the palaeoheat flux Q(t).

Computation of Ctid[j] (equation [1]) requires knowledge of the temperature history T(t,z) of the corresponding point. T(t,z) is determined by heat flux Q(t) and the thermal conductivity of the sediments K(z) that is a function of depth (z):

$$T(t,z) \; = \; T0 \; + \; Q(t) \int_{0}^{z} 1/K(z)\,dz \qquad [3]$$

Here, T0 is the temperature at surface (z=0). K(z) can be estimated from the thermal conductivities of the grain matrix K_m, pore water K_w, and porosity $\phi(z)$ which is determined by the decompaction model (surface porosity ϕ_0, compaction factor c):

$$K(z) \; = \; K_m (K_w/K_m)^{\phi(z)} \qquad [4]$$

$$\phi(z) \; = \; \phi_0 \exp(-cz) \qquad [5]$$

Given the parameters for decompaction of sediments and the thermal conductivity data for grain matrix and pore water, we can easily integrate the equations [3], [4], [5] along the burial path (Figure 1) to calculate the temperature T(t,z) from a pre-assumed heat flux Q(t). Substituting the T(t,z) in equation [1] and integrating again along the burial path, using the given activation energy E and the frequency factor A for the specified thermal indicator, the Ctid[j] is obtained. These calculations are repeated for all depth points (j) using the same pre-assumed heat flux

Q(t). Q(t) is then adjusted until all computed Ctid[j] best fit the observed Mtid[j]. Numerical solutions for searching an appropriate Q(t) to minimise the function [2] are given in the Appendix.

The uniqueness of the solution for heat flux can be demonstrated by Figure 1. The value of the thermal indicator at point 5 is a function of TTI5:

$$TID5 = F (TTI5) \qquad [6]$$
$$TTI5 = \Delta TTI4 + \Delta TTI3 + \Delta TTI2 + \Delta TTI1 + \Delta TTI0 \qquad [7]$$

It is clear that we can arrange different sets of $\Delta TTI4$ ... $\Delta TTI0$ to keep the same TTI5 that is known when the value at point 5 was given. This is the case where the solution is not unique. However, if we also have the value for point 4:

$$TTI4 = \Delta TTI3 + \Delta TTI2 + \Delta TTI1 + \Delta TTI0 \qquad [8]$$

By combining equations [7] and [8], $\Delta TTI4$ is fixed:

$$TTI5 - TTI4 = \Delta TTI4 \qquad [9]$$

Consequently, the observed values of points 1, 2, 3, 4, and 5 will therefore fix $\Delta TTI1$, $\Delta TTI2$, $\Delta TTI3$, and $\Delta TTI4$.

The boundary condition of this method includes palaeo-surface temperature T0 and present-day heat flux Q0, which can be obtained from various methods, such as down-hole temperature measurements or published data. When these boundary conditions are determined, it seems reasonable that the heat flux reconstructed from such a method is unique. The sensitivity of the boundary condition will be discussed later.

This method provides a direct determination of palaeoheat flux for the purpose of fitting the observed TID curve. The result Q(t) is not subject to any time-dependent function, i.e. the variation in palaeoheat flux with geological time is free. Specifying a function for Q(t) is also possible by this method an in this case, the coefficients of the function will be optimised. Heat flux functions are discussed further below.

The time-temperature integral on the right hand side of equation [1] may be composed of parallel reactions. For instance, the vitrinite reflectance model was considered to be composed of 20 parallel chemical reactions (8). In this case, the activation energies (E_i) and frequency factors (A_i) of each reaction will determine the transformation ratio of the thermal indicator, of which the TID value is a function (ex. $\%R_0 = \exp(-1.6+3.7F)$, where F is the transformation ratio, 8). This kind of activation energy distribution model describes the geological reaction of a thermal indicator more completely than a single reaction model (ex. 9). In addition, their kinetic parameters contribute towards a lower uncertainty for the reconstructed palaeoheat flux. See uncertainty discussion later in this paper.

Reconstruction of palaeoheat flux is based on the observed values of the thermal indicator at present-day depths. The density of sampling points will therefore determine the quality of the reconstructed palaeoheat flux. It is not necessary to have the observed points at discrete stratigraphic boundaries. Generally, the less the time interval between two points, the more precise is the control on the heat flux.

It should be mentioned that the method assumes a constant heat flux with depth. With this assumption, the variation of formation temperatures with depth at a particular time is related to the heat flux value at this time. Only one independent variable (heat flux) is to be determined for all burial depths of a particular time sequence. This makes the solution unique. One can't optimise the temperature history $T(t,z)$ directly since there is no constraint that can be made on the temperatures along the burial line of a present-day depth, or on the temperature relationship with depth. There is no unique solution for palaeo-temperatures without adding the heat flux.

Heat Flux Functions

Little is known about the variation of palaeoheat flux with geological time. Variation occurs even within a single sedimentary basin, and is geologically determined. Basement heat flow, background radioactivity of sediment, and fluid flow are the main controlling factors determining the pattern of palaeoheat flux in a specified area (*10*). A discussion of this is beyond the scope of this paper. However, thermal indicators do record the time-temperature history of sediments, and in principle it should be possible to obtain the thermal history of a sedimentary basin from this type of information.

The method proposed allows us to specify any form of functions for $Q(t)$ such as, most probably, constant, linear, exponential or polynomial heat flux variation (in these cases, we find the coefficients in a function to calculate heat flux and temperature), and even free variation of heat flux with time, i.e., where the variation of heat flux with geological time does not follow any time-dependent function. Our experiences have shown that the smallest minimum value of the objective function [2] can be achieved by such a free variation, and the heat flux reconstructed is probably more reasonable than constraining a function with time.

Dependencies of heat flux through geological time determine what kind of variables are optimized. For a free variation, we search heat flux ($Q1$, $Q2$, $Q3$, ...) directly. Otherwise we find coefficients C_0, C_1, C_2 of a function to calculate heat flux:

Constant:	$Q(t) = Q_0$
Linear:	$Q(t) = Q_0 + C_0 t$
Exponential:	$Q(t) = Q_0 \exp(C_0 t + C_1 \sin(\pi t / t_{max}) + C_2 \sin(2\pi t / t_{max}) + ...)$
Polynomial:	$Q(t) = Q_0 + C_0 t + C_1 t^2 + C_2 t^3 + ...$

where $Q(t)$ is the heat flux at time (t), Q_0 is the present-day heat flux, and t_{max} is the maximum age of strata (*3,5,11*).

It is not known which of these functions is the best. The most advantageous feature of this method is that it allows specification of any form of heat flux function with geological time. One may establish a function that fits most of wells in a sedimentary basin by trying various dependencies of heat flux on time. By doing this, a two-dimensional palaeoheat flux and thermal history reconstruction for a sedimentary basin is possible.

Uncertainty

It is not possible to derive a quantitative expression showing the uncertainty of reconstructed heat flux. Generally it is related to ① the kinetic model of the thermal indicator, ② the transformation ratio, ③ the precision of observed value, and ④ the lithology type.

Of the existing kinetic models for thermal indicators (ex. *6-9, 12-16*), it is considered better to select the activation energy distribution model than the single reaction model, since the uncertainty of kinetic parameters is distributed in each of the parallel reactions that determine the transformation ratio, of which the ultimate maturity indicator is a function (*6-8, 14*). Single reaction models (*9, 12-13*) always complete the reaction within a narrow maturity range. This limits the model itself in fitting the actual variation trend of a maturity index (*4-5*). These maturity index trends can be approached quite well by the distributed activation energy models (*8, 15-16*), besides that the distribution model describes the reaction of chemical compounds of the thermal indicator more completely in general (*7*).

The transformation ratio of the chemical reactions related to a thermal indicator is essential for determining the maturity level through a function (*6-8, 14*). The trend of variation of the transformation ratio usually follows a sigmoidal curve due to the integral form of exp(-E/RT)dt in the kinetic equation. This means only the transformation ratio within the slope interval (the middle part of the sigmoidal curve) where significant ratio changes occur upon heating will sensitively control the thermal indicator's ultimate value, and subsequently the heat flux within this maturity range. This can be further demonstrated by the fact that the heat flow in younger sediments will make a large contribution to the maturity indicator's value at great depth (Figures 2, 3), owing to pre-heating during the past. Figures 2, 3 show the influence of palaeo- and present-day heat flux on thermal maturation, calculated on a model-well using EASY%R_o kinetics (*8*). It is clearly demonstrated here that present-day heat flow is more important in determining the maturity than palaeoheat flow. This means that the uncertainty at higher maturity will make less contribution to the reconstructed heat flow at the same age. As a result, it allows more uncertainty for the observed value at higher transformation ratios or in more mature samples.

Uncertainty is also related to the lithology type of sediments. With a high thermal conductivity of the grain matrix, e.g. in sandstones, the variation of heat flow will result in significant changes of formation temperature. Inversely, the change of maturity level in this lithological type does not reflect much variation of the heat flux, although it does represent the temperature variation, i.e. the higher the thermal conductivity of sediments, the lower the uncertainty of heat flux reconstructed.

The same rule applies to the boundary conditions of this method mentioned previously. Sensitivity tests of the boundary conditions are illustrated on Figures 4, 5, which show the reconstructed heat flux of a model-well using a linear function and EASY%R_o kinetics (*8*). Limited variation of palaeo-surface temperature does not greatly affect the reconstructed heat flow. As shown on Figure 4, changing the palaeo-surface temperature from 0°C to 10°C (a typical range) results in almost the same heat flow history. However, present-day heat flow is very sensitive in determining the maturity level and hence the curvature and value of the palaeoheat flux (Figure 5). A high present-day heat flow will result in an unreasonable reconstructed palaeoheat flow. One should therefore be aware of this boundary condition. Present-day heat flux can be obtained from published data or down-hole temperature measurements. Where

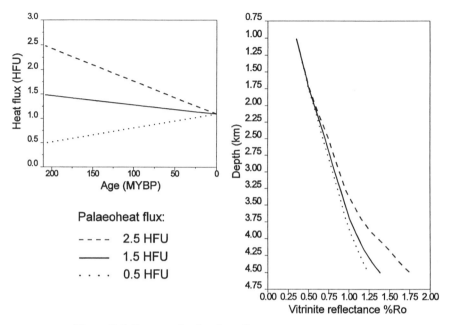

Figure 2. Influence of palaeoheat flux on thermal maturation.

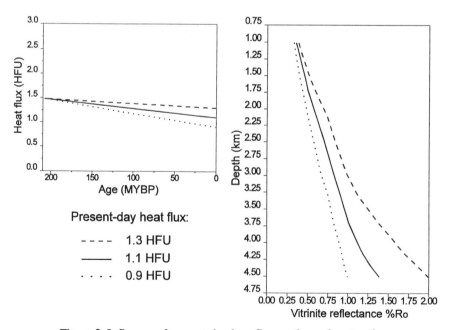

Figure 3. Influence of present-day heat flux on thermal maturation.

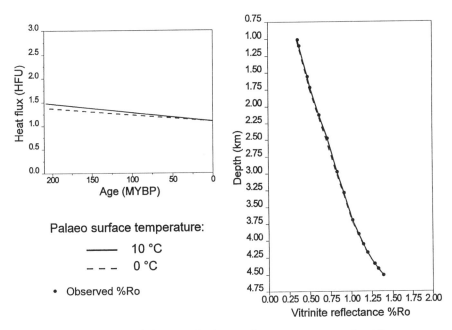

Figure 4. Sensitivity test of palaeo surface temperature on heat flux reconstruction.

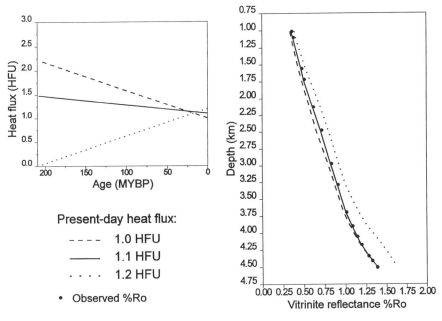

Figure 5. Sensitivity test of present-day heat flux on heat flux reconstruction.

there is no data, a function of constant heat flow with time can be used which may provide the range of present-day heat flux values.

Thermal Indicators and Working Ranges

The quality of reconstructed palaeoheat flux depends not only on the number and location of the measured values but also on the type of thermal indicator. It is considered that for validity the thermal indicator used in heat flux reconstruction should have significant changes of value within the observed maturity range (see uncertainty discussion). Figure 6 compares some of the existing thermal indicators with kinetic models calculated on a model-well. The kinetics used for preparing Figure 6 are: isomerisation (9) of steranes (C-20) and hopanes (C-22), aromatisation (13) of steroids (Aro), clay conversion (15) of smectite to illite (I/S), and vitrinite reflectance (8). There are approximately three maturation ranges reflecting significant value changes for three groups of thermal indicators: clay minerals (I/S), biomarkers (C-20, C-22, Aro), and vitrinite (%R_o), these undergoing maximum change at different maturity ranges.

The clay mineral conversion of smectite to illite starts at relatively low temperature and stops at approximately oil generation window maturity (15). The values from I/S will of course sensitively constrain the heat flow during early diagenesis (younger sediments). There is virtually no change in I/S due to variations in the palaeoheat flow at maturities higher than 0.6 %R_o. Another inorganic maturity indicator, which works over approximately the same temperature range as smectite/illite is that of fission track data.

Significant isomerisation and aromatisation of biomarkers takes place at maturities higher than that of smectite/illite conversion, approximately between early maturities and peak oil generation. The maturity indices from biomarkers are considered to be valid in modelling the palaeoheat flow at the same maturity range.

Unlike clay mineral transformation and biomarker isomerisation, the vitrinite reflectance increases continuously from immature to over-mature level. However, it is most effective at values of >0.4%R_o, partially due to the inaccuracy of measurement at low maturity. In addition, different kinetic models (6, 8) give different starting values of %R_o, however, they are consistent with each other when maturity increases.

Of the existing thermal indicators, vitrinite reflectance has the widest working range, and is valid in controlling palaeoheat flux from immature to over mature sediments. Heat flux of geologically younger sediments may be more accurately reconstructed from I/S due to significant change of percent illite layers in mixed illite/smectite clay during this early diagenesis of sediments.

Examples

Using Vitrinite Reflectance. Well A from the North Sea is mainly composed of shales and silty shales down to the total depth of 4.25 km, corresponding to a sedimentary age of 208 ma. The observed vitrinite reflectance data (from Geolab Nor A/S) are plotted on Figure 7b with measurement uncertainty ranges (solid circles with error bars). Reconstruction of palaeoheat flux was started by setting a palaeo-surface temperature of 10°C and a present-day heat flux of 1.16 HFU, using a chemical

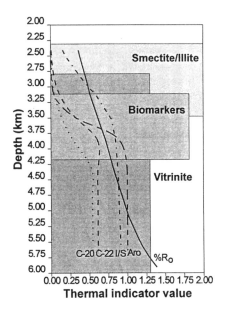

Figure 6. Comparison of various thermal indicators' maturation curves.

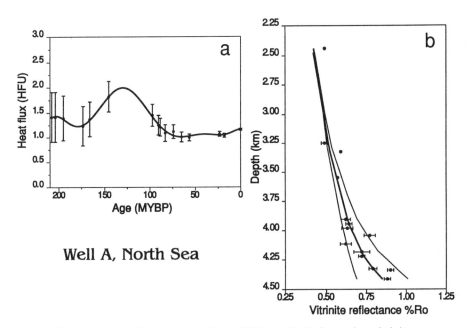

Figure 7. Heat flux reconstruction of Well A, North Sea, using vitrinite reflectance.

kinetics model (7) where $\%R_o$ is calculated from the transformation ratios of the four groups of parallel reactions describing the chemical transformation of vitrinite to H_2O, CO_2, CH_4, and CH_n. The optimized heat flow history (free variation) is plotted on Figure 7a with error bars representing the uncertainty boundary of the $\%R_o$ profile shown in Figure 7b (solid lines). The left line of Figure 7b corresponds to the lower boundary of heat flux shown on Figure 7a and the right line to the upper boundary. As seen in the figure, the uncertainty of the reconstructed heat flux is smaller in the younger sediments. The maximum uncertainty of ±0.5 HFU occurs in the deepest/oldest sediments, with a maturity of 0.88 $\%R_o$. One can expect greater variation in heat flux when the maturity at the base of the well becomes higher. The dominant lithologies of shale and silty shale with thermal conductivities of around 5.5 mcal/cm/sec/°C in this well cause the relatively large uncertainty of the reconstructed heat flow history. Palaeo-temperature calculated from the reconstructed heat flux is shown in Figure 8, with a maximum temperature of 140°C in the basal sediments of the well. The curvature of Figure 7a shows a maximum of heat flux around 130 ma, which is consistent with the start of cooling in North Sea region (17) from geophysical method of the McKenzie model (1).

Using Smectite/Illite. The smectite to illite conversion is considered to be valid in constraining the heat flow of younger sediments. Figure 9b shows the observed percent illite layers in smectite/illite clays (<0.1 μm, solid circles with error bars) from Gulf Coast Well 6 (18). Decompaction of this well was based on the stratigraphic data (18) and the typical shale/clay compaction factor and thermal conductivity data averaged from the North Sea (unpublished basin modelling default data library). Only the points between 2km to 4km were selected for reconstructing the heat flow history for this well using a distributed activation energy model (15). The lower boundary of reconstructed heat flux (free variation, Figure 9a) corresponds to the left computed line of I/S (Figure 9b, solid lines), and the upper to the right. A single reaction model (dashed line in Figure 9b) with multi- reaction order (12) can not provide a satisfactory fit since the curvature is different from the observed data.

Summary

The proposed method is a one-dimensional heat flux reconstruction using the kinetics of a thermal indicator. It provides a direct search of palaeoheat flux for the purpose of fitting the observed values of a thermal indicator at present-day depths. The most advantageous feature of the method is that it allows specification of any form of heat flux function with geological time, including free variation.

The boundary conditions of this method are palaeo-surface temperature T0 and present-day heat flux, which is more important in determining the maturity than the palaeoheat flux. Calibration of present-day heat flux against down-hole temperature measurement is recommended.

Optimising constant heat flow is computationally fastest and is one way to estimate the range of present-day heat flux if it is unknown. However, the smallest minimum value of objective function can be easily reached by specifying a free variation.

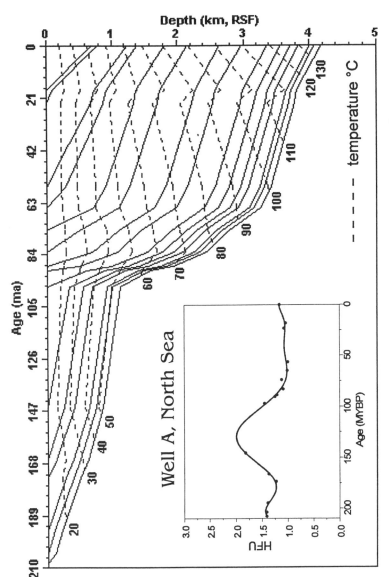

Figure 8. Temperature history for Well A, North Sea, calculated from the reconstructed heat flux shown on Figure 7.

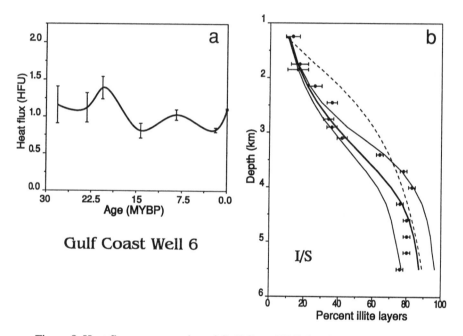

Figure 9. Heat flux reconstruction of Gulf Coast Well 6, using parallel reaction model of smectite to illite conversion.

Uncertainty is considered to be related to ① the kinetic model of thermal indicator, ② the transformation ratio, ③ the precision of the observed value, and ④ the lithology type. Generally, the higher the maturity, the more uncertain the palaeoheat flux.

The quality of reconstructed palaeoheat flux depends not only on the sampling points of measured values but also on the type of thermal indicators since they work best over different maturity ranges. It is better to use different types of maturity indicators to reconstruct the heat flow at different ages, or alternatively mixing them in one well if possible. Once some knowledge of palaeoheat flux is available, one can fix parts of the palaeoheat flux and let other parts be optimized. This is helpful in improving the quality of reconstructed thermal history.

Appendix

We convert the multidimensional minimisation problem of equation [2] to optimisation problem without constraints by taking the absolute value of heat flux since using the negative value of $Q(t)$ is incorrect. Many algorithms have been proposed for solving this kind of problem (*19-21*). We select three efficient routines, DFPDIF, POWELL, and POLYHEDRON, that use only the objective function, since it is not easy to construct derivatives for the objective function that usually involves integrals of Arrhenius equation when dealing with kinetics. Algorithms using only the objective function have the advantage of short preparation time and easy user-interface programming.

DFPDIF is a variant of the Davidon-Fletcher-Powell (DFP) algorithm that is one of the most advanced variable metric methods. DFPDIF uses difference instead of first derivative to avoid constructing a derivative. In this case, the variable metric method is still faster than those using only the function, especially when increasing the dimension. In our case DFPDIF uses Stewart's modification (*19*) of the DFP algorithm. It calls parabolic interpolation for one-dimensional search. POWELL is one of the direction set methods. It is not necessary to calculate derivatives and is a rapid algorithm. In our case POWELL uses Sargent's modification (*20*) on the original Powell algorithm, and calls parabolic interpolation for one-dimensional search. POLYHEDRON is a multidimensional optimisation without calculation of the derivative. The principle of searching the minimum value of function $f(x)$, where x is a vector of dimension n, can be briefly described as follows (*20*).

1. Construct a polyhedron of n+1 points in the n-dimensional space E^n, within which, any n points are not in the subspace of n-1 dimension of E^n.
2. Calculate the value of objective function $f(x)$ at each point.
3. Select the point where the objective function $f(x)$ has the maximum value. Shine upon the other points to find a point that has less value of $f(x)$. Construct a new polyhedron.
4. Repeat 2-3 to improve the polyhedron until no better point can be obtained.
5. Compress or expand the polyhedron and repeat 2-4 until the points converge to the minimum value of $f(x)$.

Of the three none-constrained algorithms provided, POLYHEDRON method is the best in numerical stability. It does not require the objective function $f(x)$ to be continuous. The only disadvantage is the relatively slow speed. POLYHEDRON can be used in most basin modelling problems.

There is no perfect optimisation algorithm. It is helpful to try more than one method and more than one set of starting values of independent variables to get global minimum and reasonable results. Heat flux reconstruction using the POLYHEDRON algorithm is not sensitive to initial values. DFPDIF and POWELL algorithms may give slightly different heat flux curves with different initial values when specifying a non-linear function with time. Virtually nothing is known about finding a global minimum in general (21).

Acknowledgements

This is a part of the project "Petroleum Generation and Migration in North Sea, An Experimental and Numerical Approach" financed by Geolab Nor A/S. We appreciate Peter B. Hall and Ian L. Ferriday for their review of the manuscript and improvement of English usage.

Literature Cited

(1) McKenzie, D. *Earth and Planetary Science Letters.* **1978**, *40*, 25-32.
(2) Curiale, J. A.; Larter, S. R.; Sweeney, R. E.; Bromley, B. W. In *Thermal History of Sedimentary Basins: Methods and Case Histories;* Naeser, N. D.; McCulloh, T. H., Ed.; ISBN 0-387-96702-8; Springer-Verlag: New York, 1989; pp53-72.
(3) Lerche, I.; Yarzab, R. F.; Kendall, C. G. St. C. *The American Association of Petroleum Geologists Bulletin.* **1984**, *68*, 1704-1717.
(4) Armagnac, C.; Kendall, C. G. St. C.; Kuo, C.; Lerche, I.; Pantano, J. *Journal of Geochemical Exploration.* **1988**, *30*, 1-28.
(5) Marzi, R.; Rullkotter, J.; Perriman, W. S. *Advances in Organic Geochemistry 1989. Organic Geochemistry.* **1990**, *16*, 91-102.
(6) Larter, S. R. *Marine and Petroleum Geology.* **1988**, *5*, 194-204.
(7) Burnham, A. K.; Sweeney, J. J. *Geochimica et Cosmochimica Acta.* **1989**, *53*, 2649-2657.
(8) Sweeney, J. J.; Burnham, A. K. *The American Association of Petroleum Geologists Bulletin.* **1990**, *74*, 1559-1570.
(9) Mackenzie, A. S.; McKenzie, D. *Geol. Mag.* **1983**, *120*, 417-470.
(10) Allen, P. A.; Allen, J. R. *Basin Analysis: Principles and Applications*; ISBN 0-632-02423-2; Blackwell Scientific Publications: Oxford, London, 1990; pp282-305.
(11) Liu, J.; Lerche, I.; Bissada, K. K.; Lacey, J. *Organic Geochemistry.* **1992**, *18*, 385-396.
(12) Pytte, A. M. *The kinetics of the smectite to illite reaction in contact metamorphic shales*; M.A.thesis, Dartmouth College, Hanover, New Hampshire, 1982; 78pp.
(13) Abbott, G. D.; Maxwell, J. R. *Advances in Organic Geochemistry 1987. Organic Geochemistry.* **1988**, *13*, 881-885.
(14) Senftle, J. T.; Larter, S. R.; Bromley, B. W.; Brown, J. H. *Organic*

Geochemistry. **1986**, *9*, 345-350.
(15) Wei, H.; Roaldset, E.; Bjorøy, M. EAPG 5th Conference, Extended Abstracts, paper P551, Stavanger, 7-11 June, 1993.
(16) Ritter, U.; Aareskjold, K.; Schou, L. *Organic Geochemistry.* **1993**, *20*, 511-520.
(17) Hermanrud, C.; Eggen, S.; Larsen, M. In *Generation, accumulation, and production of Europe's hydrocarbons; Special Publication of the Europen Association of Petroleum Geoscientists*; Spencer, A. M., Ed.; Oxford University Press: Oxford, Vol. 1; pp.65-73.
(18) Hower, J.; Eslinger, E. V.; Hower, M. E.; Perry, E. A. *Geological Society of American Bulletin.* **1976**, *87*, 725-737.
(19) Stewart, G.W. *J. ACM.* **1967**, *14*, 72-83.
(20) *FORTRAN Algorithm Assemblage*; Liu, D. G.; Fi, J. G.; Yu, Y. J.; Li, G. Y., Ed.; ISBN 7-118-00297-6; National Defence Industry Press: Beijing, 1983; Vol.2; pp242-391.
(21) Press, W. H.; Flannery, B. P.; Teukolsky, S. A.; Vetterling, W. T. *Numerical Recipes in C*; ISBN 0-521-35465-X; Cambridge University Press: Cambridge, 1988; pp290-352.

RECEIVED April 15, 1994

Author Index

Affiliation Index

Subject Index

A

Alberta, Canada, reflectance suppression in cretaceous coals, 93–108

Algal-derived corpohuminite-like bodies, *See* Corpohuminite from Canadian Paleozoic source rocks

American Society for Testing and Materials
coal classification, 27
measurement standards, vitrinite reflectance, 29,30*t*
round-robin, vitrinite reflectance data, 32,33*f*,34
training kit, vitrinite reflectance data, 32

Anthracenes, distribution in vitrinites, 142–146

Apatite fission track analysis
age vs. temperature, 250,251*f*
description, 249
limitations, 217

Apatite fission track thermochronology integrated with vitrinite reflectance
annealing of fission tracks vs. evolution of vitrinite reflectance, 252,253*f*
Burnt Timber Thrust, 261–265
complementary advantages, 256–259
examples, 254,256
kinetic-model evaluation, 250,252–255
predictive advantages, 259–261,262*f*

Applications, vitrinite reflectance, 26–29

Aromatic compound distribution by class, flash pyrolysis–GC–MS of Lower Kittanning vitrinites, 138,140–142

Artificial thermal maturation, evolution of vitrinite ultrafine structures, 194–204

B

Basin modeling, application of vitrinite reflectance, 15–17

Beneficiation of coal, measurement using vitrinite reflectance, 29

Bituminous coal, transformation of vitrinite, 112–135

Boron content role in vitrinite reflectance suppression in cretaceous coals from Alberta, Canada, 106

Burial heating, comparison of kinetic models and vitrinite reflectance geothermometry, 219,223–227

Burnt Timber Thrust, Southwest Alberta, Canada, movement using apatite fission track thermochronology integrated with vitrinite reflectance
regional setting, 261,262*f*,263
thermal history interpretation, 263–265

Butane yield, role of pressure in coal pyrolysis, 170

C

C_2–C_5 gases, pressure effect on coal pyrolysis, 187,189

Calibration standards, vitrinite reflectance, 34

Canadian Paleozoic source rocks, corpohuminite, 52–62

Carbon dioxide yield, pressure effect during coal pyrolysis, 174,176*f*,189

Carbon-isotope analysis, role of pressure in coal pyrolysis, 174,177*t*,178*f*

Carbon isotope–pyrolysis method for thermal maturity determination of kerogen, *See* Pyrolysis–carbon-isotope method for thermal maturity determination of kerogen

Cell structure in vitrinite, remanent, *See* Remanent cell structure in vitrinite

Chemical characterization, Spanish Jurassic jet, 81

Chemical structural composition of coal, 113

Clusters, definition, 136

Production: Beth Harder & Charlotte McNaughton
Indexing: Deborah H. Steiner
Acquisition: Rhonda Bitterli
Cover design: Amy Hayes

Printed and bound by Maple Press, York, PA

Highlights from ACS Books

Good Laboratory Practice Standards: Applications for Field and Laboratory Studies
Edited by Willa Y. Garner, Maureen S. Barge, and James P. Ussary
ACS Professional Reference Book; 572 pp; clothbound ISBN 0–8412–2192–8

Silent Spring Revisited
Edited by Gino J. Marco, Robert M. Hollingworth, and William Durham
214 pp; clothbound ISBN 0–8412–0980–4; paperback ISBN 0–8412–0981–2

The Microkinetics of Heterogeneous Catalysis
By James A. Dumesic, Dale F. Rudd, Luis M. Aparicio, James E. Rekoske,
and Andrés A. Treviño
ACS Professional Reference Book; 316 pp; clothbound ISBN 0–8412–2214–2

Helping Your Child Learn Science
By Nancy Paulu with Margery Martin; Illustrated by Margaret Scott
58 pp; paperback ISBN 0–8412–2626–1

Handbook of Chemical Property Estimation Methods
By Warren J. Lyman, William F. Reehl, and David H. Rosenblatt
960 pp; clothbound ISBN 0–8412–1761–0

Understanding Chemical Patents: A Guide for the Inventor
By John T. Maynard and Howard M. Peters
184 pp; clothbound ISBN 0–8412–1997–4; paperback ISBN 0–8412–1998–2

Spectroscopy of Polymers
By Jack L. Koenig
ACS Professional Reference Book; 328 pp;
clothbound ISBN 0–8412–1904–4; paperback ISBN 0–8412–1924–9

Harnessing Biotechnology for the 21st Century
Edited by Michael R. Ladisch and Arindam Bose
Conference Proceedings Series; 612 pp;
clothbound ISBN 0–8412–2477–3

From Caveman to Chemist: Circumstances and Achievements
By Hugh W. Salzberg
300 pp; clothbound ISBN 0–8412–1786–6; paperback ISBN 0–8412–1787–4

The Green Flame: Surviving Government Secrecy
By Andrew Dequasie
300 pp; clothbound ISBN 0–8412–1857–9

For further information and a free catalog of ACS books, contact:
American Chemical Society
Distribution Office, Department 225
1155 16th Street, NW, Washington, DC 20036
Telephone 800–227–5558

Bestsellers from ACS Books

The ACS Style Guide: A Manual for Authors and Editors
Edited by Janet S. Dodd
264 pp; clothbound ISBN 0–8412–0917–0; paperback ISBN 0–8412–0943–X

The Basics of Technical Communicating
By B. Edward Cain
ACS Professional Reference Book; 198 pp;
clothbound ISBN 0–8412–1451–4; paperback ISBN 0–8412–1452–2

Chemical Activities (student and teacher editions)
By Christie L. Borgford and Lee R. Summerlin
330 pp; spiralbound ISBN 0–8412–1417–4; teacher ed. ISBN 0–8412–1416–6

Chemical Demonstrations: A Sourcebook for Teachers,
Volumes 1 and 2, Second Edition
Volume 1 by Lee R. Summerlin and James L. Ealy, Jr.;
Vol. 1, 198 pp; spiralbound ISBN 0–8412–1481–6;
Volume 2 by Lee R. Summerlin, Christie L. Borgford, and Julie B. Ealy
Vol. 2, 234 pp; spiralbound ISBN 0–8412–1535–9

Chemistry and Crime: From Sherlock Holmes to Today's Courtroom
Edited by Samuel M. Gerber
135 pp; clothbound ISBN 0–8412–0784–4; paperback ISBN 0–8412–0785–2

Writing the Laboratory Notebook
By Howard M. Kanare
145 pp; clothbound ISBN 0–8412–0906–5; paperback ISBN 0–8412–0933–2

Developing a Chemical Hygiene Plan
By Jay A. Young, Warren K. Kingsley, and George H. Wahl, Jr.
paperback ISBN 0–8412–1876–5

Introduction to Microwave Sample Preparation: Theory and Practice
Edited by H. M. Kingston and Lois B. Jassie
263 pp; clothbound ISBN 0–8412–1450–6

Principles of Environmental Sampling
Edited by Lawrence H. Keith
ACS Professional Reference Book; 458 pp;
clothbound ISBN 0–8412–1173–6; paperback ISBN 0–8412–1437–9

Biotechnology and Materials Science: Chemistry for the Future
Edited by Mary L. Good (Jacqueline K. Barton, Associate Editor)
135 pp; clothbound ISBN 0–8412–1472–7; paperback ISBN 0–8412–1473–5

For further information and a free catalog of ACS books, contact:
American Chemical Society
Distribution Office, Department 225
1155 16th Street, NW, Washington, DC 20036
Telephone 800–227–5558